AF564550

A Fully Illustrated Handbook on Diagnostic Cytology of Dog and Cat

NIPA® GENX ELECTRONIC RESOURCES & SOLUTIONS P. LTD.
New Delhi-110 034

About the Authors

Dr. C. Balachandran was born on 28.07.1957 in Mannargudi town of Tiruvarur District, Tamil Nadu. He obtained B.V.Sc., M.V.Sc. and Ph.D. (Veterinary Pathology) degrees from Madras Veterinary College. He is Charter Member and Diplomat of Indian College of Veterinary Pathologists (ICVP). He is also PG Diploma holder in Agricultural Journalism-PGDAJ and Ethnoveterinary Practices PGEVP. He started his career at Namakkal in 1980 and promoted as Associate Professor in 1988 and Professor in 1996. He served as Vice-Chancellor of Tamil Nadu Veterinary and Animal Sciences University (TANUVAS) for three years from 09-04-2018. He also served as Registrar of TANUVAS and as the Dean, Madras Veterinary College and Faculty Dean of Veterinary and Animal Sciences, TANUVAS. He has published about 400 scientific articles including 73 articles in international journals. He has worked on skin tumours, mammary tumours, lymphadenopathies, and also pathology of liver, spleen, kidney, stomach, intestine and colon mostly on cytology and histopathology. He was instrumental in introducing cytological diagnosis of diseases in animals. Other areas of research interests are avian diseases, cancer biology and indigenous medicinal effects. He has about 90+ awards to his credit. Notable Awards: University gold medal for the best thesis in avian diseases, Fellow of FAO, IAVP, National Academy of Veterinary Sciences and Tamil Nadu State Scientist Award.

Dr. N. Pazhanivel, M.V.Sc., Ph.D., M.A. (Public Administration), PGDEVP., Diplomate ICVP., FASAW, FIAVP, ICMR DHR IF, and Professor of Pathology, Madras Veterinary College, Chennai-600007. His area of research includes oncology, avian oncogenic viruses, avian pathology, cytopathology, histopathology, fish histopathology and toxicopathology. He has published 322 research articles (International -65; National-257). He has published 64 popular articles. He has presented 323 research papers (International-91; National- 232). He has received 61 awards. He has published 3 books and 56 book chapters. He has completed 6 projects and currently 7 projects are handled. He has completed Post-Doctoral Fellowship Research Training at Pennsylvania State University, USA in 2020. He visited to Liverpool University, United Kingdom for educational training program under ICAR- NAHEP /IDP –TANUVAS during January to February, 2023. He is acting as the Editor, The Indian Veterinary Journal from 2022 onwards. He acted as the Secretary, Indian College of Veterinary Pathologists (ICVP) from 2020 and now acting as Registrar, Indian College of Veterinary Pathologists (ICVP) from 2023 onwards.

Dr. S.Vairamuthu, Professor and Head Centralised Clinical Laboratory, Madras Veterinary College, Chennai-600 007 Tamil Nadu, India. He has 33 years of professional experience including 16 years of experience in Clinical Laboratory of Madras Veterinary College. Teaching UG students for 25 Years, Post graduate for 20 years and guided 15 post graduate students and 3 PhD scholars. Expertise in Interpretation of haematology and serum biochemistry results. Screening of cytological smears for neoplastic, inflammatory conditions and blood smears for blood parasites and blood pictures. Diagnostic Cytology – screening of FNAC smears / impression smears from cutaneous and internal masses for diagnosis of tumours. Visted University of Maryland, Maryland USA short term training (3 months) in animal diseases diagnosis (molecular aspects) in Virginia Maryland regional College of Veterinary Medicine. Visited University of Bari, Italy training program on Educational Technology on Veterinary Medicine at College of Veterinary Medicine ICAR- NAHEP /IDP -TANUVAS. Visited South Africa and Nairobi-Kenya -BBSRC, UK-TANUVAS collaborative Project. Published more than 120 research papers including 35 in foreign journals. He has been received more than 20 awards including seven in national level.

A Fully Illustrated Handbook on Diagnostic Cytology of Dog and Cat

C. Balachandran
Professor of Veterinary Pathology
Former Vice Chancellor
Tamil Nadu Veterinary and Animal Sciences University
Madhavaram Milk Colony
Chennai – 600 051. Tamil Nadu, India

N. Pazhanivel
Professor of Veterinary Pathology
Madras Veterinary College
Chennai-600 007, Tamil Nadu

S. Vairamuthu
Professor and Head
Centralised Clinical Laboratory
Madras Veterinary College
Chennai-600 007, Tamil Nadu

NIPA® GENX ELECTRONIC RESOURCES & SOLUTIONS P. LTD.
New Delhi-110 034

NIPA® GENX ELECTRONIC RESOURCES & SOLUTIONS P. LTD.

101,103, Vikas Surya Plaza, CU Block
L.S.C.Market, Pitam Pura, New Delhi-110 034
Ph : +91 11 27341616, 27341717, 27341718
E-mail: newindiapublishingagency@gmail.com
www: www.nipabooks.com

For customer assistance, please contact
Phone: + 91-11-27 34 17 17
Fax: + 91-11-27 34 16 16

Print ISBN: 978-93-58876-80-2

ebook ISBN: 978-93-58878-10-3

Composed and Designed by NIPA®.

Preface

Cytology, the examination of cells to identify different conditions, has become an integral part of clinical pathology as it provides quick diagnosis of certain conditions which is helpful to the clinician and practicing Veterinarians.

"A Fully Illustrated Handbook on Diagnostic Cytology of Dog and Cat" provides information forming an important source for cytological diagnosis of various inflammatory and non-inflammatory conditions of the dog and cat. This book gives basic information on cytological description of conditions to the veterinarians, clinicians, researchers and students who begin their carrier on this subject.

Cytology is nothing but examination of cells from living animals. It is less invasive technique than tissue biopsies and, in some cases, it can provide a rapid diagnosis compared to histopathology. In conditions were clinical examination, radiographic evaluation, and laboratory data may not be sufficient to confirm diseases, cytology comes to help in confirming the disease processes or conditions. This includes collection of appropriate representative cells, preparation, processing of the specimen and interpretation which will give successful cytological diagnosis. This book act as one of the diagnostic and educational tool to the veterinarians, faculties and students.

Dr C. Balachandran worked at Centralised Clinical Laboratory, Madras Veterinary College, Chennai (1999-2006), He could initiate and establish cytological diagnosis of diseases in animals mostly in dogs. We have started working on surface swellings/conditions and lymphadenopathies initially and later working on internal organ pathology (spleen, liver, stomach, intestine etc. in dogs). Few works were also being carried out in cats and cattle. From these collections, we thought of writing this illustrated diagnostic cytology book.

Dr. S.Vairamuthu, has worked at Centralised Clinical Laboratory, Madras Veterinary College, Chennai from 2007 to 2024 also used cytological diagnosis of diseases in animals.

Dr.N.Pazhanivel has also been worked in cytological diagnosis of diseases from materials received from Madras Veterinary College Clinics and also from practicing Veterinarians to Referral Laboratory for Diagnostic Cytology and Histopathology at Department of Veterinary Pathology, Madras Veterinary College, Chennai.

I am indebted to Dr. B. Murali Manohar, then Professor and Head, Department of Veterinary Pathology, Madras Veterinary College, Chennai who motivated and helped in establishing cytological diagnosis of diseases in animals.

Many students were assigned work in the topics on neoplasms i.e. skin, mammary gland and lymphadenopathies in dogs and molecular studies were made later. I thank all the students in general, specially Dr. K. Krithiga and Dr. M. Thangapandiyan for sharing some pictures.

We are thankful to Tamil Nadu Veterinary and Animal Sciences University, Chennai for the facilities provided and support.

Authors

Contents

1

Introduction

Cytology is the study of cells from fluid or tissue samples prepared directly on glass slides and stained for microscopic evaluation. It is also described as examination of cells without regard to the tissue's architectural details.

It is less invasive than tissue biopsies and, in some cases, it can provide a rapid diagnosis that would otherwise take several days if done by histopathology.

When is the cytology needed?

The diagnostic methods of clinical examination, radiographic evaluation, and laboratory data suggested that a diagnosis that cannot be confirmed by these methods. In that situation, cytology is very much helpful to confirm the disease processes or conditions.

At present, tumour diagnosis is needed to use less invasive method without using expensive instrument (Sangha *et al*., 2011). The success of cytological diagnosis mainly depends upon collection of appropriate representative cells, preparation, processing of the specimen and interpretation of smears and all these procedures are interdependent (Roszel, 1981; Lumsden and Baker, 2000).

The arrangement of neoplastic cells within tissue is critical to determine the diagnosis of many types of tumours, in evaluating surgical margins and in establishing whether a tumour is benign or malignant (Stirtzinger, 1988). The role of cytology as a diagnostic tool in Veterinary Medicine is continuously and constantly developing and expanding. Over the last two decades, it has become established as a reliable quick method of obtaining a tissur for diagnosis in a minimally invasive way.

Nowadays, the clinicians have increased their uses of the diagnostic techniques and cytopathologists are becoming more experienced with a wider variety of lesions, the wide variety of tissues samples handled, the spectrum of disease processes identified by cytology, reliability and precision of the diagnosis must be increased (Meinkoth and Cowell, 2002).

Besides, the cytology is considered as a screening diagnostic tool and various lesions can be classified as inflammatory/non-neoplastic, hyperplastic, or neoplastic/non-inflammatory. For neoplastic process, an experienced cytologist can definitively diagnose several specific neoplasms, make a tentative diagnosis of neoplasia for many types of tumours, identify sites of tumour metastasis, and monitor tumour regrowth following anticancer therapy. Information gained from cytology may be useful in establishing a diagnosis, determining a prognosis, and formulating a diagnostic or therapeutic plan.

2

Biopsy and Cytological Methods

A judicious use of cytology will be beneficial to the practicing veterinarians as diagnosis is available within a shortest period of time and can decide on proper treatment e.g. TVT can be successfully treated medically without surgical intervention.

Diagnostic Approach in Clinical Set Up

- Clinical investigation
- Clinical pathology
- Necropsy
- Biopsy
- Cytology
- Histopathology
- Electron microscopy
- Fluorescent antibody techniques
- Immunohistochemistry
- Genomic techniques
- Microarray
- Proteomics
- Metabolomics etc.

Biopsy

Biopsy is collection of materials from living animal for examination and diagnosis of diseases in animals.

Biopsy Methods

- Fine needle aspiration biopsy-FNAB
- Excisional biopsy

- Incisional biopsy
- Punch biopsy
- Irrigation of hollow organs
- Abdominal/thoraco/pericardiocentesis
- Scrapings
- Tape (Sellotape) stripings
- Swab smears
- Brushings
- Curettage
- Pressure massage

Among these various methods listed, fine needle aspiration biopsy aspiration is the most preferred method that is easy to perform at various set up of veterinary practice.

Cytology

Examination of cells without regard to the tissue's architectural details

Cytology specimens can be obtained by various biopsy techniques.

George Papanicoloau is claimed as "Father of clinical cytology".

Exfoliative Cytology

Exfoliative cytology is the examination of cancer cells that are already shed for anaplasia and origin of tumour e.g. ethmoid tumours in cattle and dogs. It is also routinely used in stages in estrus cycle in cycling female dogs.

Cytology-Indications

Materials can be obtained from variety of tissues, body cavities etc.

- Soft tissue masses: Cutaneous & subcutaneous lesions, enlarged lymph nodes, Intra-thoracic/abdominal masses/fluids/ Body cavity effusions
- Suspected splenic or hepatic pathology
- Prostatic aspirates
- Bone marrow evaluation

When

- Clinical examination
- Radiographic evaluation
- Laboratory data suggest a diagnosis that cannot be confirmed by these methods

Contraindications-Many

Important ones are

- Haemorrhagic tendencies
- Inexperienced clinician
- Malignancy-Needle biopsy-Metastasis-Remote; But do not hesitate to do FNAB
- Perform biopsy to confirm or eliminate clinical diagnosis
- Establish final diagnosis or evaluation of therapy

Advantages-Cytology as sampling especially FNAB

- Quick
- Easy
- Inexpensive
- Less invasive than surgical biopsy
- No general anaesthesia
- Light sedation
- Short processing time
- Results available in minutes
- Fewer complications

Diagnostic Cytology-Limitations

- No tissue architecture
- Tissue origin
- Biologic behaviour
- Grading
- Degree of tissue infiltration

- Presence and absence of necrosis
- Not representative of the lesion
- Small mobile or very firm lesions don't yield sufficient materials
- Mesenchymal – Concurrent HP required
- Obese-Peri lesion fat
- Large lymphoma - LN – Necrotic - Non - diagnostic
- Impression/scraping-Ulcer-Superficial cells
- Secondary inflammation/infection

Artefacts

- Poor smear preparation-Crushing causes smudged cells-"bare" nuclei; nuclear & cytoplasmic vacuolation/strands
- Dirty slides
- Water artefacts
- Starch granules- Gloves
- Ultrasound gel
- Tissue paper or plant fibres
- Pollen grains
- Stain precipitates

To differentiate artefacts/contaminants from intracellular structures focus and unfocus. Artefacts after unfocusing appear as refractile dots.

Fine Needle Aspiration Biopsy

"Stop collecting sample if blood appears in syringe"

Fine needle aspiration biopsy (FNAB) is one of the most common biopsy method by which cytological materials can be obtained from living animals. The method can be used to diagnose inflammatory and neoplastic conditions. Neoplasms arising from different cells can also be diagnosed. FNAB can obtain cells from almost any part of the body. Safe and often painless method in proper hands; As long as the needle is correctly positioned in the target tissues, the aspiration usually generates an abundance of highly diagnostic cells. The exact location of the needle point and its progress after insertion in the deeper tissue is usually monitored by palpation.

- Employed to identify cells
- Differentiate inflammation and neoplasia

Advantages

- Less time consuming
- Single procedure – multiple site sampling
- Consecutive sampling at intervals
- Diagnosis may eliminate surgery
- Specimens also used for HP

Limits

- Does not always permit positive diagnosis-support or contradict a tentative clinical diagnosis
- Not for study of structural changes

FNAB General Procedure

- Clean the area aseptically
- Anaesthetize the area as required
- Hold the mass firmly in hand
- Use 3 to 20 mL syringe with 21 to 25 gauge needle depends on soft to hard mass/swelling
- Insert needle into the mass, apply negative pressure and move the needle in various directions while aspirating
- Release negative pressure, Withdraw needle out
- Dislodge needle and air drawn into the barrel
- Push the contents of the needle on clean glass slides
- Prepare smears

May be suction or non-suction type based on tissue involved

Continuous suction-Done as above

Intermittent suction

- More suitable for smaller lesions, where it is not possible to advance or redirect the needle without exiting the mass

- 23G needle; 5 mL syringe
- Withdraw and release the plunger several times with the needle inserted in the mass
- Release the suction and remove the needle from the mass
- Disconnect the needle from the syringe
- Push the contents of the needle on clean glass slides
- Prepare smears

Non-suction

i. Needle alone technique

- Causes less damage to fragile cells e.g. lymphoid cells
- Minimizes blood contamination
- Useful for aspirating highly vascular masses and lymph nodes
- Obtaining ultrasound guided samples of lesions within the body cavities
- Insert needle without the syringe into lesion after disinfection of the skin
- Rapidly move the needle back and forth at least 10 times before withdrawal
- Attach to a syringe with about 3-5 mL of air and expel the aspirated material onto a slide
- Prepare smears

ii. Needle with syringe technique

- Insert needle with syringe having about 2-3 mL air into the lesion
- Rapidly move the needle and syringe back and forth at least 10 times before withdrawal
- Remove the syringe with needle and expel the aspirate on to the slide
- Prepare smears

"Metastasis is remote with FNAB sampling; Do not hesitate to collect samples by FNAB" **Preparation of Slides**

- Fluid – Smears – Many techniques
- Small fragments – Squash

- Large tissue fragment – Imprints/HP
- Culture – Swirl the needle in medium before fixation for HP/slide preparation

Blood film technique

- Useful for fluid aspirate – but not those with low cellularity
- Expel aspirate from the syringe downward to one end of the slide
- Keeping the spreader slide with one end missing at 20-45° angle draw back until it comes into contact the sample drop
- Advance the spreader slide gently for wards to create a smear with a feather edge after the sample has spread along the interface of the two slides

Line concentration technique

- Modified blood film technique
- Useful for hypocellular fluid aspirates e.g. cystic masses or samples heavily contaminated with blood
- The sample is smeared as for a blood film but after the spreader has been advanced about 2/3 way along the bottom slide it is lifted abruptly upwards.
- Has the effect of concentrating cells along the line at the end of the smear

Squash preparations

- Expel the aspirate onto the centre of the slide
- Place a second spreader slide horizontally and at right angles
- Spread the sample, taking care not too much downward pressure to avoid rupturing of cells
- Quickly draw the slide smoothly across the bottom of the slide
- In most cases the weight of the spreader slide is sufficient to spread the cells
- The smear produced on the underside of the spreader slide that is examined under microscope

Processing Fluid Samples

- Collect fluid aspirates into EDTA and plain tube if microbiology required
- Centrifuge Low cellularity fluids – Transudates, peritoneal fluid and CSF samples at low speed – 1000-1500 rpm
- Resuspend the sediment with a drop of supernatant and smear – Some artefacts
- Cytospin centrifuge deposits cells directly on a predefined area of the slide

Punch/Needle Core Biopsy

- Indicated when structural relationship within the lesions are considered essential for evaluation
- Use biopsy needle specifically designed for the purpose – Variety available
- Tru-cut needle 14G – Three handed operation
- Vet-core needle 14-20G: Single handed operation
- Both: Single use – Ethylene oxide or cold sterilization
- Greater risk of post biopsy complication than FNAB
- On removal place the tissue in fixative or prepare impression smears

Incisional Biopsy

- Small portion of a lesion surgically removed is utilized

Grab Sampling

- Small tissue samples from mucosal surface (Oro-pharynx, nasal sinus, bladder, stomach)
- Fibreoptic endoscopy have an integral, centrally hinged, cup biopsy instrument capable of recovering superficial samples
- Hinge-action ring forceps are also used for proctological sampling
- All other procedures are same as excisional biopsy.

Tape Striping

- Choose the sampling site based on the gross appearance of the affected area

- Clipping the hair surrounding the lesion to enhance organism yield
- Press a Sellophane tape (5.5 cm length x 2.5 cm width) to the skin lesion for 2-3 seconds
- Apply to the slide
- Useful in malasseziasis

Excisional Biopsy

- Advantageous over FNAB
- Preferred since both treatment and diagnosis are incorporated
- Large tissue removed – HP/cytology
- Include adjacent normal tissue along with the lesion
- Incise tissue for imprint smears-cytology
- Disadvantage – Time consuming and costly
- Use acceptable surgical procedure

Impression

- Ulcerated surface
- Cut surface of excised or PM specimens
- Carefully blot the cut surface of the tissue with absorbent material/paper towel to remove the blood and other body fluid
- Make impression smears by gently touching flat cut surface
- Make several imprints on the same slide
- Air dry or wet fix
- Remaining portions sliced into thin sections and fixed for HP

Impression - Disadvantages

- It collects relatively a few cells
- Cells collected may not necessarily be representative of the lesion under investigation. e.g. Ulcerated lesion are often secondarily inflammed or infected which may result in only inflammatory cells of interest or may induce in the cells of interest dysplastic changes that mimic those associated with neoplasia

- Limited use in superficial neoplasia
- Useful for intraoperaive confirmation of neoplasia for biopsy samples or excised tissue

Scrapings

- Touch imprints from solid tissues acellular
- Harvesting cells from lesions unlikely to yield large number of cells on FNAB -Mesenchymal
- So, scrape smears are prepared
- Scrape cut surface carefully with a scalpel blade
- Discard the first scraping contaminated with the blood and again scrape carefully
- Spread materials in a thin layer/squash on glass slides
- Air dry or wet fix

Swab Smears

- Fistulous tracts, vagina or ear canal
- Moisten swab with normal saline to minimise cell damage
- Rub against the surface of the lesion or vaginal wall
- Gently roll on to a clean glass slide

Brushings

- Very soft friable specimens e.g. splenic haemangiosarcoma
- Cells transferred from he cut surface of the excised tissue with a fine camel hair brush
- GIT by passing a brush through the biopsy channel of an endoscope
- Conjunctiva, respiratory tract, vaginal sampling

3

Fixation

Fixation can be done by either wet fixation or dry fixation. If staining is to be delayed for a few days, cell preservation enhanced by fixation. Fix the slide within 30-60 seconds.

Wet Fixation

- If staining is to be delayed for a few days cell preservation enhanced
- Fix slide within 30-60 seconds / Spray
- 95% ethanol/absolute isopropanol /methyl alcohol for 20 - 30 min
- RBC lysis improves exam clumps of cells

Dry Fixation

- By rapidly agitating in air

4

Cytological Stains

Different types of Romanowsky type of stains are used for staining the cytological smears. It will give excellent cytoplasmic and nuclear details and it also stains bacteria. The following stains are commonly used. e.g., Wrights, Giemsa, Leishman, Wright-Giemsa, May-Grunwald-Giemsa, Leishman-Giemsa, Haematoxylin and Eosin, Diffquik stains, Trichrome stains, Papanicolaou or Sano modification.

Staining Methods

Leishman-Giemsa

- Cover the smear with LG stain for a minute
- Dilute with tap water/Distilled water/ phosphate buffered saline pH 6.8-7.2
- Allow to stain for about 20 min
- Wash in water

Wright

- Flood air dried smears with 1:1 Wright's stain and distilled water
- Wash smears after 3-4 min

Giemsa

- Flood the air died slide with freshly prepared 1:10 Giemsa stain and distilled water and allow it to remain for 40 minute.
- Wash in running tap water
- Air dry the smear

Leishman

- Flood the air died slide with Leishman stain and allow it to remain for 1 minute.

- Add double the quantity of distilled water and allow it to remain for 15-20 minutes.
- Wash in running tap water
- Air dry the smear

Wright-Giemsa

- Flood air dried smears with Wright's stain for 3-4 min
- Wash smears and flood with 1:10 Giemsa and distilled water for 15-20 min
- Wash in running tap water

May-Grunwald-Giemsa Stain

- Flood air dried smears with 1:1 May- Grunwald and distilled water for 3-4 min
- Wash smears and flood with 1:10 Giemsa and distilled water for 15-20 min
- Washed in running tap water

Haematoxylin and Eosin (H&E)

- Wet fixed smears kept in Harri's haematoxylin for 20 min
- Wash in running water
- Dip in 1% acid alcohol, wash immediately
- Keep in running water for 5-10 min for "bluing"
- Stain with 1% eosin for 1-2 min
- Wash, dry and dip in xylol
- Mount on DPX mountant with clean cover slip

Trichrome Stains

- Wet fixed smear
- Rehydrate air-dried cells for 30 sec in saline and wet fixation before staining
- Superior nuclear and nucleolar detail

- Cytoplasm stains weakly
- Useful for thick smears or
- Those contaminated heavily with blood
- Time consuming
- Not generally practical for a veterinary laboratory

New Methylene Blue

- Rarely used as cytological stain
- Aqueous stain that allows immediate examination of a smear after application to air-dried cells
- Place a drop of stain on the smear
- Examine under microscope

Other Stains

Toluidine Blue

Mast cell granules in poorly differentiated mast cell tumour

Fontana Stain

Malignant melanoma with very few melanin pigment granules

Periodic Acid-Schiff

Fungal hyphae and fungal spores

We generally use Leishman-Giemsa cocktail stain. For academic interest we have used other individual and cocktail stains.

Staining Characters

- Romanowsky stains and combinations with Giemsa stain nucleus purplish and cytoplasm bluish
- Haematoxylin & eosin stain the nucleus bluish and cytoplasm eosinophilic

5

Cytological Evaluation of Slides

Location is important

Since, one should recognize whether cells belong to that area and / or not

Basic aim

- To establish an aetiological and/or morphological diagnosis in order to obtain a more accurate prognosis

Salient Points in Cytological Examination

- Cytology is a joint effort between the clinician and cytologist
- Adequate history, clinical signs and methods of collection should be intimated
- Interpretation suffers if done in a vacuum of information
- Cytology is an rapid, inexpensive technique in tumour diagnosis

Check the Following

- Quality of the specimen good to evaluate
- Cells normal to the anatomical site

Ideal Smear

- Intact
- Well stained cells should have a clearly evident demarcation between the nucleus and cytoplasm

Differences Between Fine Needle Aspiration Biopsy and Exfoliative Cytology

- In FNAB cells alive at the time of fixation and do not show the physiologic changes of ageing and death
- These young cells are more active and frequently mitotic thus likely to be over diagnosed, if they show some traumatic structural distortions.

- Where as in exfoliative cytology, the shed cells are examined

Microscopic Evaluation

- Place a cover slip cover the smear that will improve visibility and clarity of cells up to 40x.
- Scan at low magnification-4x, 10x objective to identify cellularity or areas with different staining characteristics, crystals, foreign bodies and parasites
- 20x to assess cellularity and cellular composition i.e. different cell types inflammatory, epithelial cells
- 40x dry and then 100x oil immersion for evaluation of cell morphology in greater detail
- Low power to view cellularity
- Type of cells
- Distribution of cells
- Nature of lesion
- Discrete / Individual
- Sheets
- Clusters
- Papillary
- High power to view cellular details

Assess Whether the Condition is

- Inflammatory/reactive or non-inflammatory
- Inflammatory: Predominant cell types; aetiological agent present i.e. bacteria/fungus
- Non-inflammatory: Neoplastic/Non-neoplastic, dysplastic, if neoplastic origin of cells present-Benign or malignant

Cytological Feature in Hyperplasia

- Inflammatory reactions often result in hyperplasia of surrounding tissues.
- Resemble normal cells except appear more immature.

- Large nuclei with poorly condensed chromatin and prominent nucleoli.
- Cytoplasm often basophilic.
- Constant nuclear-to-cytoplasmic ratio
- An important feature in distinguishing hyperplastic from neoplastic cells.

Cytological Features of Neoplasia

Basic important criteria

Nuclear/Nucleolar criteria: Note

- Size
- Shape
- Number
- Mitosis
- Staining character

Benign Neoplasia

- Cells small and similar size with small nuclei (To the inexperienced cytologist it is difficult to give a guide as to what constitutes 'small'. However, there are invariably some red cells scattered through the smear and these can be used as a 'measuring stick').
- The nuclei of benign cells are usually >2 times the diameter of a red cell
- Little variation in cell and nuclear size
- Nucleus: Cytoplasm (N:C) ratio low i.e. Relatively small nucleus surrounded by abundant cytoplasm
- Nuclei either do not contain nucleoli or may contain one or occasionally two, small nucleoli that are uniform in size and shape
- Similar appearance to normal tissue and to hyperplasia

Malignant Neoplasia

Non-specific cellular criteria

- Contain more cells-Hypercellular-Not a reliable criterion-Operator skill, size of needle used and type of tumour – (Mesenchymal tumours exfoliate less well than round cell tumours)

- Greater pleomorphic cells – Anisocytosis (2-3 fold) & shape – Increased larger cells (1.5 to 2 times than normal) with variable N:C ratio
- Population of cells in an abnormal location – Good indicator of neoplasia – Even if the cell population is monomorphic and nuclear criteria for malignancy are absent e.g. cutaneous lymphoma-Small well differentiated lymphocytes present

Cytoplasmic Criteria

Less definitive than nuclear criteria

- May also occur in hyperplasia/dysplasia
- However if present along with nuclear criteria for malignancy, they are supportive of a diagnosis of malignancy well defined
- Increased cytoplasmic basophilia – Dark blue cytoplasm - Increased RNA content – may be seen in any metabolically active cells e.g. hyperplastic cells, secretory glandular epithelial cells – so not necessarily indicative of neoplasia
- Cytoplasmic margins- may be less well defined
- Cytoplasm may appear vacuolated or excessively granulated in malignant cells and
- Large coalescing vacuoles or "Signet ring" cells where nucleus pushed to the periphery of the cell by a single large cytoplasmic vacuoles
- Vacuoles may be present in cells that are not usually vacuolated e.g. malignant squamous cells or sarcoma cells
- Evidence of acinus formation highly suggestive of malignancy (carcinoma)

Nuclear Criteria

- Anisokaryosis-Variation in nuclear size – 2-3 times of RBC diameter
- Variation in nuclear shape
- Nucleus may be distorted (nuclear moulding) compressed by adjacent neoplastic cells or by a second nucleus within the same neoplastic cells
- Multinucleation (more significant if the nuclei within the same cell vary in size)
- Increased no. mitotic figures (may be seen in hyperplasia or benign)

- Two or more/x40 field significant
- Atypical/asymmetrical/abnormal mitotic figures where nuclei divide in more than two directions or nuclear division is uneven is more significant and is highly suggestive of malignancy
- Increased N:C ratio except lymphoid cells very low N:C ratio (Normal - 1: 3 1:8 depending on tissue becomes < 1:2 significant) or variation in N:C ration within the same population of cells
- Single population of cells containing young immature blastic nuclei highly suggestive of malignancy especially with evidence of nuclear and nucleolar pleomorphism
- Multiple nucleoli (3-4) in one nucleus or a single very large nucleolus – RBC size
- Variation in nucleolar size, shape and number especially within the same nucleus – Extremely large, angular - fusiform, triangular or irregularly shaped nucleoli highly suggestive of malignancy
- Normal nucleoli-Round
- Coarse, ropy or reticulated or cord-like nuclear chromatin pattern. Small white spaces are seen between clumps.
- Non-neoplastic cells have smooth nuclear chromatin with a uniformly dark appearance

Diagnosis of Malignancy

- Recognition of >3 of the above nuclear criteria is regarded as strong evidence for malignancy
- Fewer than 4 criteria or subtle changes – may be benign, hyperplasia, dysplasia or malignancy
- Requires histological evaluation

Two Pit Falls – To be avoided when looking for malignancy

Bare Nuclei

- Whose cytoplasm has been disrupted during smearing – 'skipocytes'
- These nuclei will swell up and appear enlarged, with coarse nuclear chromatin and prominent nucleoli giving the appearance of malignancy
- These nuclear changes are artefactual and ignored

Tissue Cells are Accompanied by Inflammatory Cells

- Severe inflammation may cause tissue cells to become dysplastic; nuclei may enlarge, there may be variation in size of nuclei and nucleoli, basophilic cytoplasm, coarse chromatin and increased N:C ratio

May Be

- Primary inflammation with secondary dysplasia or
- Primary neoplasia which has elicited inflammatory response
- To clear doubt-biopsy and histology

Cytology Interpretation

- Based chiefly on changes in the appearance of individual cells.
- In the hands of experts, false positive results are uncommon, but false negative do occur, because of sampling errors.
- When possible, cytological diagnosis must be confirmed by biopsy methods.

Histopathology-*Gold Standard in Diagnosis*

Histopathological (HP) confirmation is essential in all cases - Definitive and morphological diagnosis. 1cm thickness tissue samples (incisional or excisional biopsies) are fixed in 10% neutral buffered formalin/formal saline in 10-20 times the volume of specimen. Routine paraffin embedded sections are prepared. Haematoxylin and Eosin (H&E) stainis usually employed.

For immunohisto chemistry (IHC), fixation time should not exceed 48 hours. Special staining techniques can also be used for identifying kind of tissues e.g. van Gieson (Fibrous tissue) and components e.g. iron (Perl's), melanin (Fontana).

Surgical Margins

Safe margins are given to eliminate chances of remnants of tumour tissue during surgery. Usually 1-2 cm margins are given depending on the situation. HP helps to rule out the presence of tumour tissues in the margins. Thereby pathologists render helping hand to the surgeon in giving away the prognosis.

Frozen Sections

An important tool in surgical pathology and performed adjacent to the surgical theatre. The results of frozen section agree with about 90% of the results of paraffin sections. The frozen sections are stained with polychrome methylene blue or with haematoxylin and eosin and graded according to their cellular differentiation. It is not superior to histopathology.

Histological Grading of Tumours for Prognosis

Based on Assessment of Morphologic Criteria Like

- Degree of cellular differentiation
- Invasiveness
- Overall cellularity
- Mitotic index and necrosis alone or in combination
- Tumours are classified as well differentiated to poorly differentiated/ low to high grade

Cytology poorly differentiates tumors, not a substitute for histopathological examination

We need to accept the "historical supremacy of routine histology" in diagnostic pathology. Histopathology is confirmatory in diagnosis of diseases with architectural details available. Modern technologies can be an important and exciting adjunct to routine pathological examination.

Jones and Fletcher (1999)

Neoplasms are cytologically classified as benign or malignant according to the following criteria of malignancy (Tyler *et ai.*,1999)

S. No.	Terminology	Description
1.	Anisocytosis	Variation in the size of cell
2.	Pleomorphism	Variation in the size and shape of the same cell type
3.	Hypercellularity	Increase in the cell exfoliation because of weakened connections between cells
4.	Anisokaryosis	Variation in the size of nucleus
5.	Macrokaryosis	Increase in size of nucleus
6.	Multinucleation	More number of nuclei in the cell
7.	Increased nuclear/ cytoplasmic ratio.	Ratio of nucleus and cytoplasm was increased
8.	Nuclear molding	Nucleus present one over another within the cell (Deformation of the nuclei)
9.	Increased mitotic figures and abnormal mitoses	More mitotic figures and abnormality in mitoses
10.	Coarse chromatin pattern	Multiple chromatin clumps as coarse chromatin
11.	Macronucleoli	Larger size of the nucleoli
12.	Angular nucleoli	Angulation of nucleoli seen
13.	Anisonucleosis	Variation in the size of nucleoli

The evaluation of cytology smear is given by algorithm

Criteria in Evaluating the Cytological Smears for Malignancy

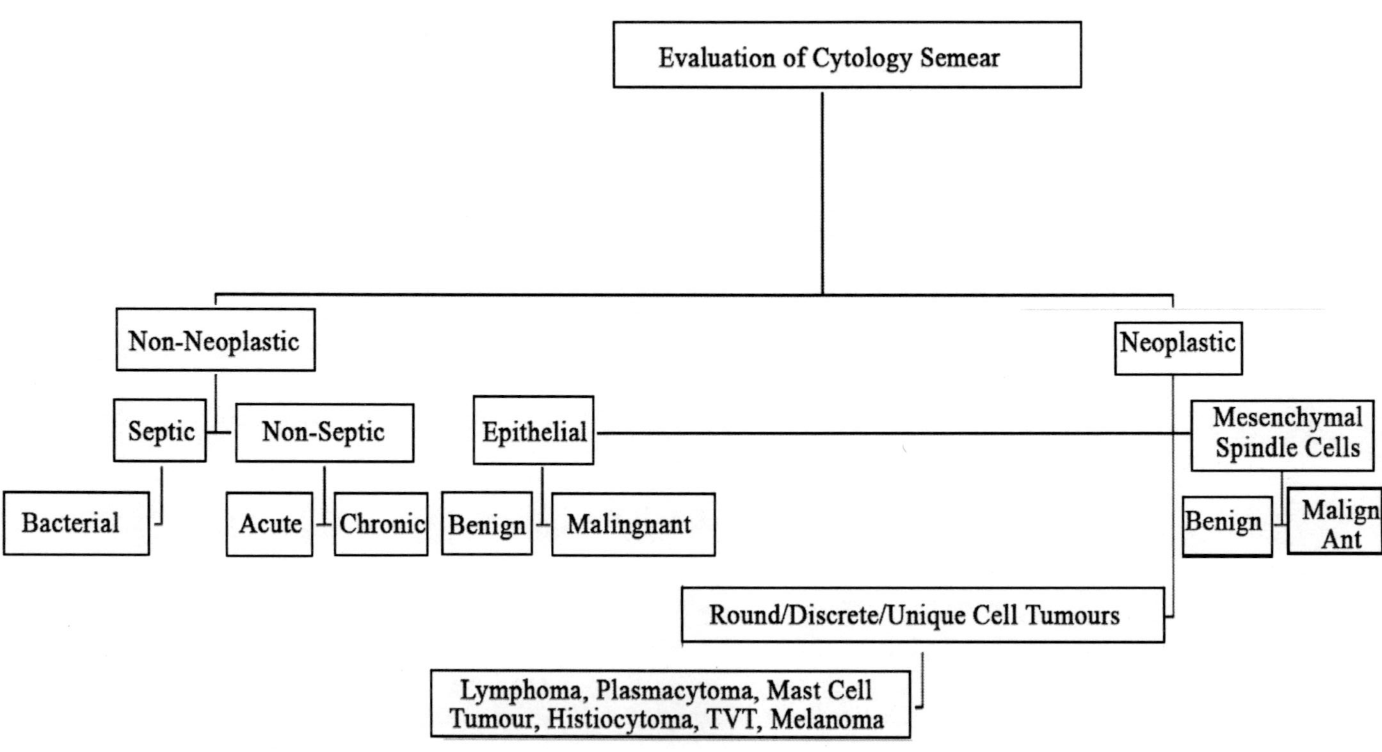

6

Ideal Smears

Good smear is well spread and tongue shaped. If forced during discharge of materials from the syringe, it is sprayed on to slide.

The ideal smears will yield intact and well stained cells (Figure) should have a clear demarcation between the nucleus and cytoplasm

Poor smear will show no cellularity or damaged cells.

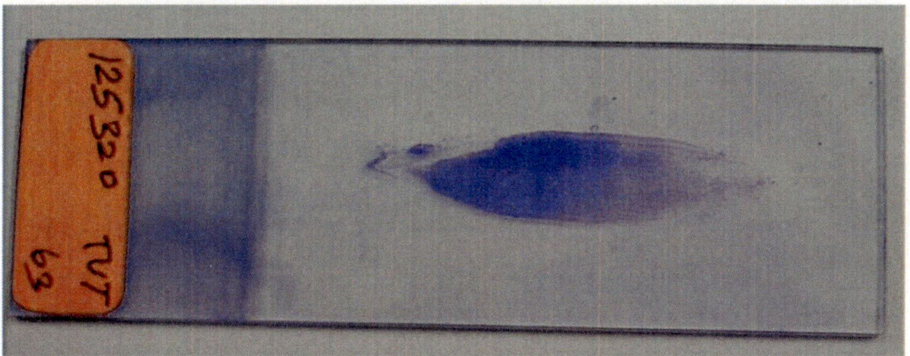

Good smear

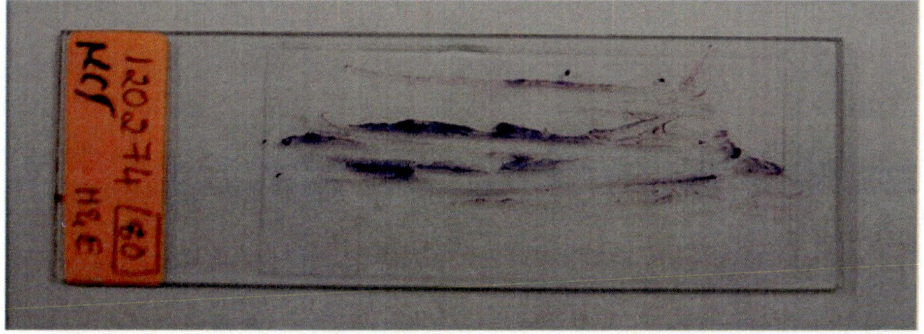

Sprayed smear

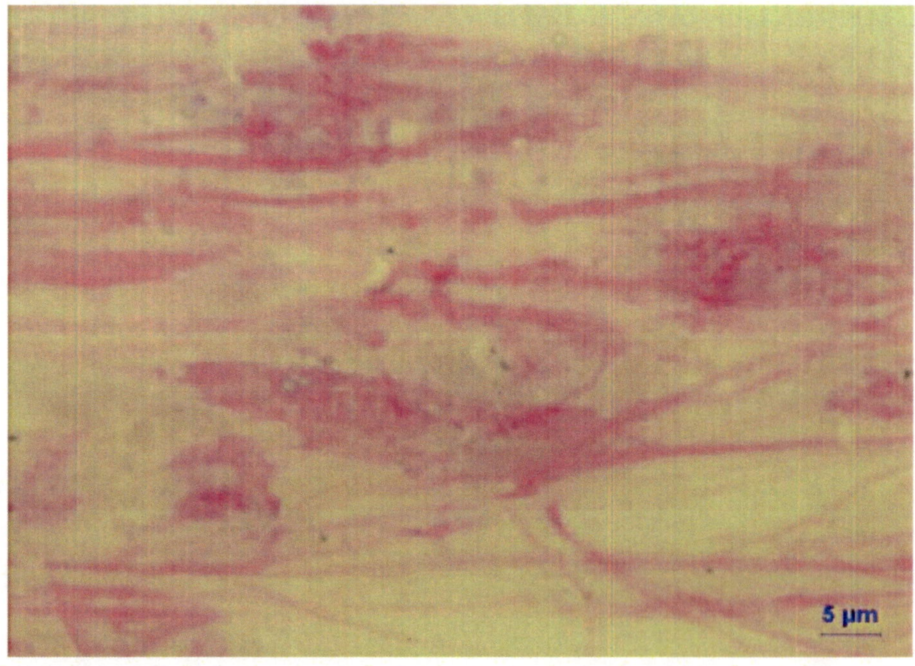

Badly prepared smear: Distorted cells without clarity of nucleus and cytoplasm

7

Cytological Diagnosis of Non-Neoplastic and Inflammatory Conditions

Answer Inflammatory or Not

- If yes it is non-neoplastic or inflammatory
- If no, consider neoplastic conditions
- Certain non-neoplastic conditions may not show inflammatory cells (Haematoma, cystic fluid, epidermal/dermal cysts)
- Inflammation or non-neoplastic condition is characterized by a mixed population of cells including neutrophils, lymphocytes, plasma cells, eosinophils, macrophages, epithelioid cells and giant cells.
- Inflammatory reactions often result in hyperplasia of surrounding tissues.
- If not inflammatory, it may be neoplastic.

Classification Based on Duration

- Acute
- Subacute
- Chronic active
- Chronic

These terms refer to different types of inflammation than to the duration of inflammatory response, since in reality, the relative percentages of inflammatory cell types do not reliably predict duration.

Classification Based on Severity of Inflammation

Assessed subjectively

- Mild
- Moderate
- Severe

Classification Based on Predominant Cell Types

- Purulent/suppurative (Neutrophils)
- Eosinophilic
- Mixed (neutrophils, lymphocytes, plasma cells and macrophages)
- Granulomatous (Mixed but predominantly MNC and macrophages)

Classification-Combination

- Acute-Purulent/suppurative or neutrophilic response
- Chronic active-Mixed inflammatory response
- Chronic-Granulomatous (Some- Giant cells and epithelioid macrophages)
- Not mutually exclusive e.g. pyogranuloma

Acute Inflammation

- >70% (some 85%) of total number of nucleated cells neutrophils
- Remaining cells mixture of monocytes, macrophages, lymphocytes and plasma cells
- Cell morphology may provide clue to the aetiology

Neutrophils

- If neutrophil is >85%, the term suppurative is used.
- Sometimes, the nucleus is essentially round, mimicking a lymphocyte, but the cells can still be identified as a neutrophil by the lobulated outline of the nucleus.
- Neutrophils may undergo several morphologic changes in tissues. Aging change is a commonly encountered phenomenon. The initial change seen is hypersegmentation of nucleus

- The nucleus of the cell swells, appearing thicker, staining more eosinophilic, and losing nuclear lobulation.

Degenerative Neutrophils

- Swelling and hyalinisation of nucleus ("Glass-like" appearance i.e. It is shiny but not refractile)
- Loss of nuclear lobulation and rupture of nuclear membrane (Karyolysis)
- Karyorrhexis (Fragmentation of nucleus) and formation of nuclear dust
- Pyknosis-Shrinkage of nuclear material into dark staining hyperchromatic masses
- Caused due to bacterial toxins (Gram negative bacteria) or irritant material

Non-degenerative Neutrophils

- Well preserved neutrophils as that of peripheral blood
- Associated with non-septic/sterile conditions e.g. immune-mediated meningitis or polyarthritis
- Absence of degenerative neutrophil not exclude bacterial aetiology e.g. Joint fluid even in confirmed septic arthritis
- Expected to improve with antibiotic therapy
- Body cavity effusions e.g. with high protein content or blood degenerate in significant delay in processing (This aging process occurs in vitro and does not necessarily indicate an infectious aetiology)

Karyolysis and Karyorrhexis

- Indicate sudden cell death
- Degenerative neutrophils showing karyolysis associated with sepsis esp. Endotoxin producing or pyogenic bacteria, e.g. pyothorax

Pyknosis

- Slow cell death, associated in the with aging of neutrophils in the inflammatory process
- Less indicative of sepsis
- Some bacterial infection do not produce marked degenerative changes e.g. *Nocardia*

Look for in neutrophils

- Examine for presence of ingested bacteria
- Presence of intracellular bacteria makes contamination after sampling less likely and
- Supports a diagnosis of bacterial infection
- Examine for presence of ingested bacteria
- Presence of intracellular bacteria makes contamination after sampling less likely and
- Supports a diagnosis of bacterial infection

Neutrophils - ulcerative Cutaneous Lesions

- Invariably covered with inflammatory cells and bacteria
- Impression smears not representative of underlying pathological process
- Prior removal of superficial exudate and cellular debris with a saline-soaked gauze swab may provide a more representative impression smear

Neutrophils in certain tumours e.g. SQCC

- Frequently elicit an intense inflammatory response
- Thorough examination of smear is required to detect neoplastic cells

Inflammation

- Induce dysplastic changes in normal cells
- Resemble neoplastic cells

Eosinophilic Inflammation

- Subcategory of acute inflammation
- >10% eosinophils
- Cytoplasm shows rust or brown granules
- Hypersensitivity to inflammation
- Presence of parasitic agent
- Feline eosinophilic granuloma
- Paraneoplastic syndrome: Mast cell tumours, lymphomas

- Less frequent in lymphomas and histiocytoma – response to cytokine release
- Eosinophilic granuloma-Spleen

Lymphocytic/Plasmacytic Inflammation

- Associated with allergic/ immune stimulation (vaccine site reaction)
- Chronic inflammation
- Rhinitis
- Lymphoplasmacytic inflammatory bowel disease
- Lymphoid cell population may be relatively mixed (heterogeneous i.e. small to medium-sized lymphocytes and plasma cells with other inflammatory cells)
- But a predominant monomorphic population of small lymphocytes without other inflammatory cells help to distinguish the lesions from lymphoma.

Chronic Inflammation

- Predominance of macrophages (These are foamy, often vacuolated and phagocytic cells)
- At least 50% macrophages & remaining admixture of neutrophils, lymphocytes and plasma cells
- Low grade irritants (Inert foreign materials)
- Fungi – *Histoplasma capsulatum*
- Bacteria - *Mycobacterium*, *Actinomyces, Nocardia*
- Resolving acute inflammatory reaction

Macrophages

- Predominance of macrophages indicate a more insidious or chronic inflammation e.g. Mycobacteria spp.
- Macrophages have an extremely varied morphology in tissues, which can be somewhat confusing.
- In some chronic inflammatory lesions, "EPITHELIOID" macrophages may be encountered. This description is applied to macrophages that

are enlarged with expansive cytoplasm that stains uniformly basophilic, giving the cell the look of an epithelial cell.

- Readily identified by size – 20-100 µm diameter
- 3-12 times diameter of RBC
- Oval or bean shaped nucleus
- Abundant pale-blue cytoplasm (Romanowsky stains) may contain numerous vacuoles and / or phagocytosed cellular materials
- Nucleus often eccentrically placed
- Lacey or reticulated chromatin pattern

Granulomatous Inflammation

- Subcategory of chronic inflammation
- Characterized by significant number of large reactive and smaller epithelioid macrophages
- Response to foreign material or persistent intracellular infectious agents
- In some cases inflammatory giant cells
- Small number of plasma cells, neutrophils and lymphocytes

Epithelioid Cells

- Modified macrophages
- Smaller macrophages
- Large polygonal shape
- Abundant basophilic cytoplasm
- Not actively phagocytic
- Resemble epithelial cells
- Seen in small clusters
- Foreign body reaction or mycobacterial infections

Inflammatory Giant Cells

- Large multinucleated cells
- Differentiate from neoplastic giant cells

- Nuclei of inflammatory giant cells uniform in size and shape (Pleomorphic in neoplastic mulinucleated cell) and arranged around periphery of the cell
- Increased no. a feature of foreign body reactions

Chronic Active Inflammation

- Mixed inflammatory response with chronic and active components
- 50-70% Neutrophils
- 30-50% mononuclear cells (Lymphocytes & Macrophages)
- Macrophages and neutrophils major cell types – Pyogranuloma
- Migrating grass horns, keratin release from epidermal inclusions/ sebaceous cyst
- Fungal infections
- Certain bacteria *Mycobacterium, Actinomyces, Nocardia*
- Panniculitis
- Lick granuloma
- Other chronic injuries

Pyogranulomatous Inflammation

- Term reserved for a population of neutrophils and epithelioid macrophages with or without multinucleate giant cells

Haematoma

Collection of blood outside the blood vessels.

It is more common in external ear.

Cytology

- Presence of erythrocytes, blood filled leococytes (non-degenerative neutrophils and macrophages) and platelets
- Haemosiderin or haematoidin crystals will be seen.

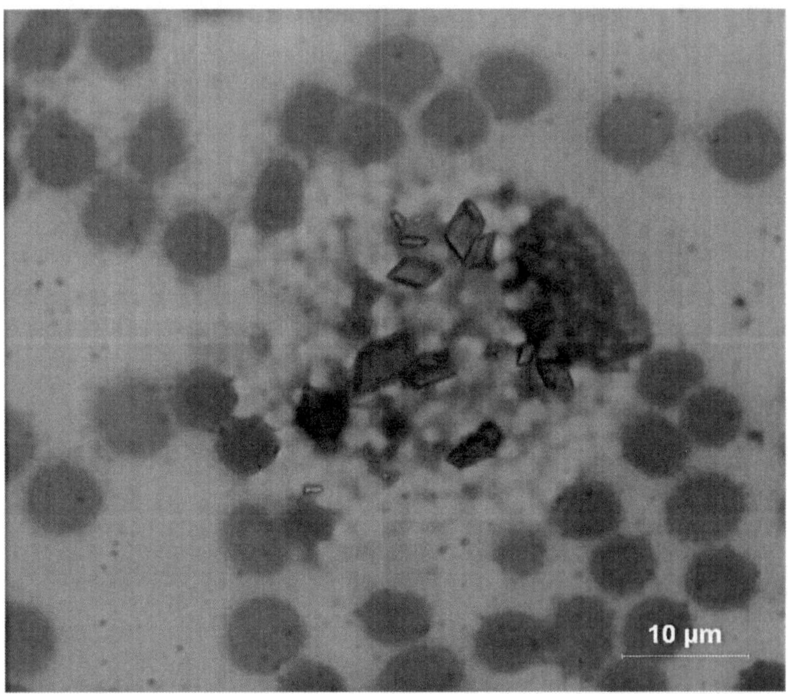

Presence of haematoidin crystals in the macrophage Leishman & Giemsa 10 µm

Cyst

- Cyst originate from either benign or malignant conditions
- Both neoplastic or non-neoplastic lesions are seen as cystic mass
- Fluid filled cyst lined by thin walled columnar or cuboidal epithelium
- Parathroid cyst, thyroid cyst, thryoglossal cyst, bronchial cleft cyst and epidermal inclusion cyst

Cytology

- Well preserved cells arranged in papillary form, dark bland nuclei with vacuolated cytoplasm
- No cellularity
- Eosinophilic fluid will be seen

Acute condition

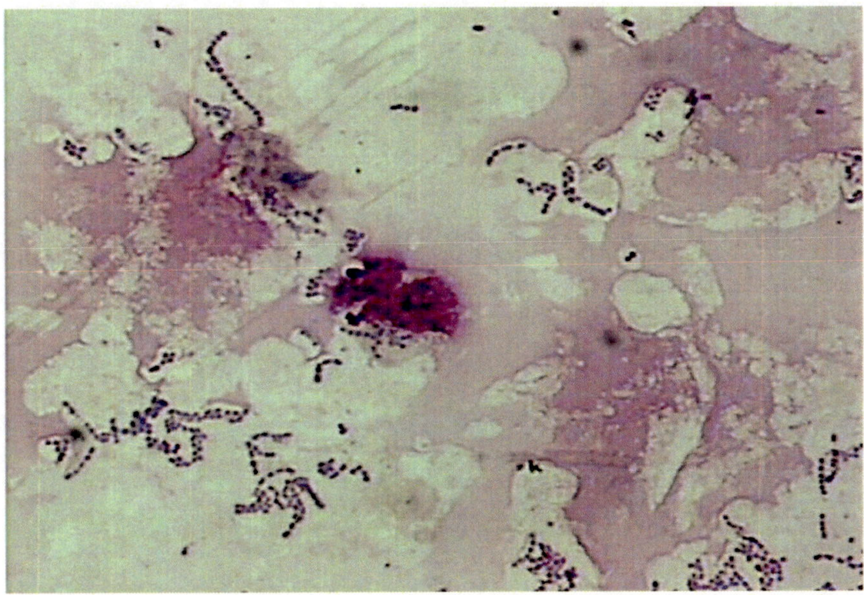

Septic condition-Abscess- Skin- Streptococcal infection- Presence of phagocytosed bacteria in the neutrophils and chains of bacteria. Leishman & Giemsa . Bar =100x

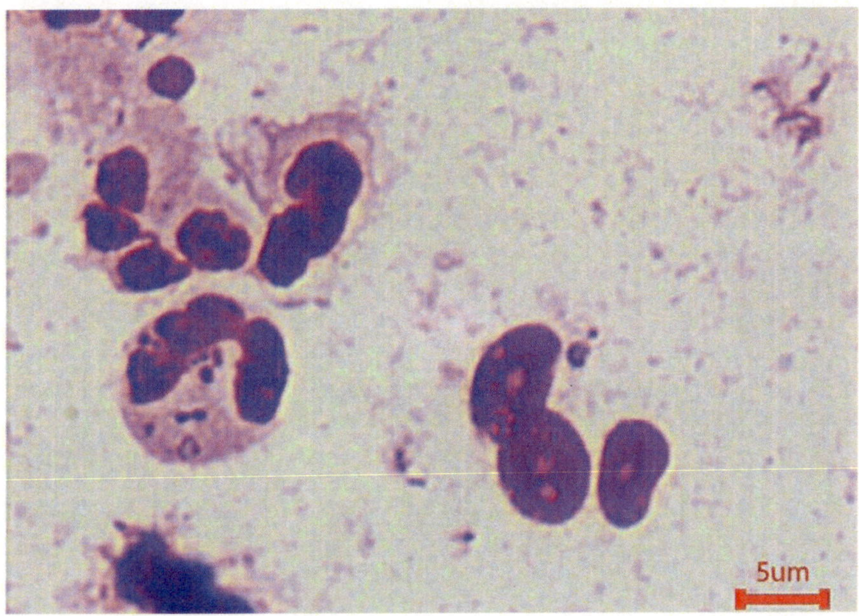

Septic inflammation- Skin- Presence of mature and immature neutrophils. Phagocytosed bacteria in the neutrophil Leishman & Giemsa Bar=5 μm

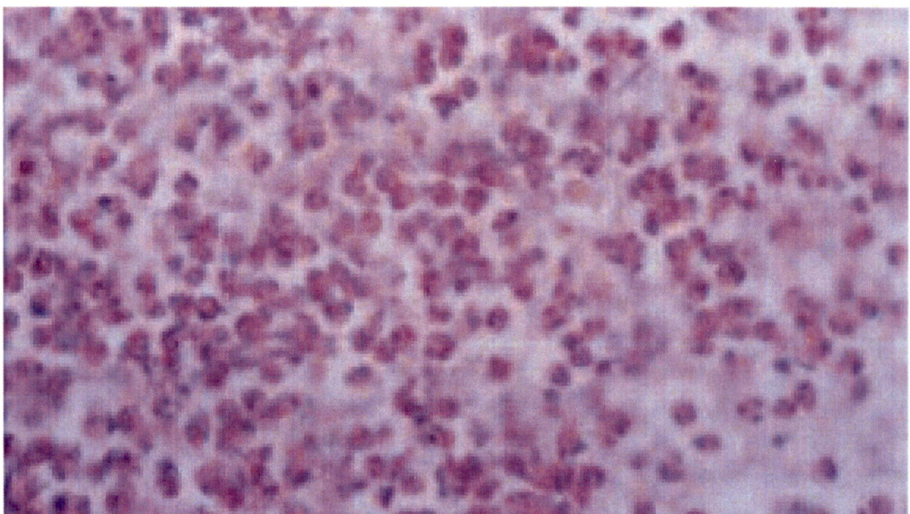

Equine – Lavage-Laryngitis – Eosinophils Leishman & Giemsa 100x

Chronic conditions

Tuberculosis

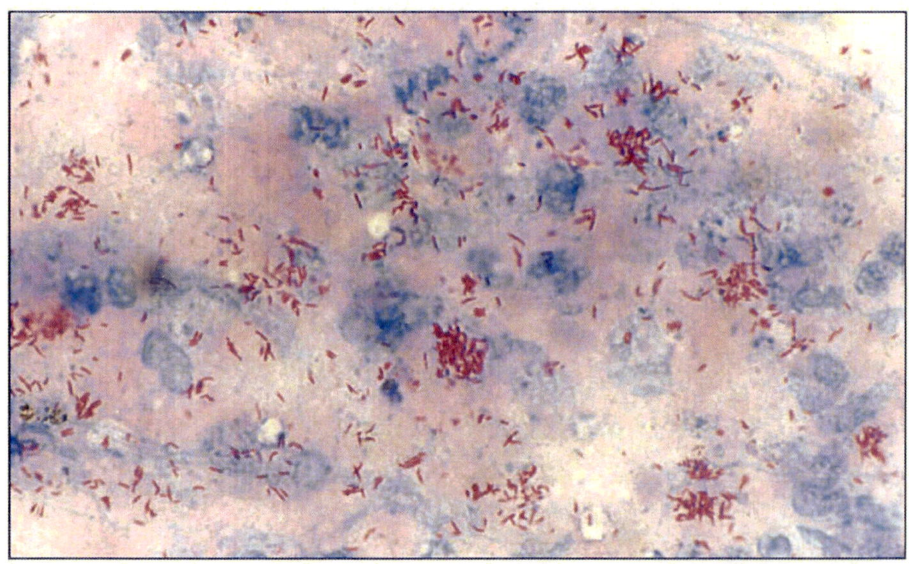

Cattle- *Mycobacterium tuberculosis*- Single to clusters of thin rods of acid fast bactreria. Ziehl-Neelsen stain 100 x

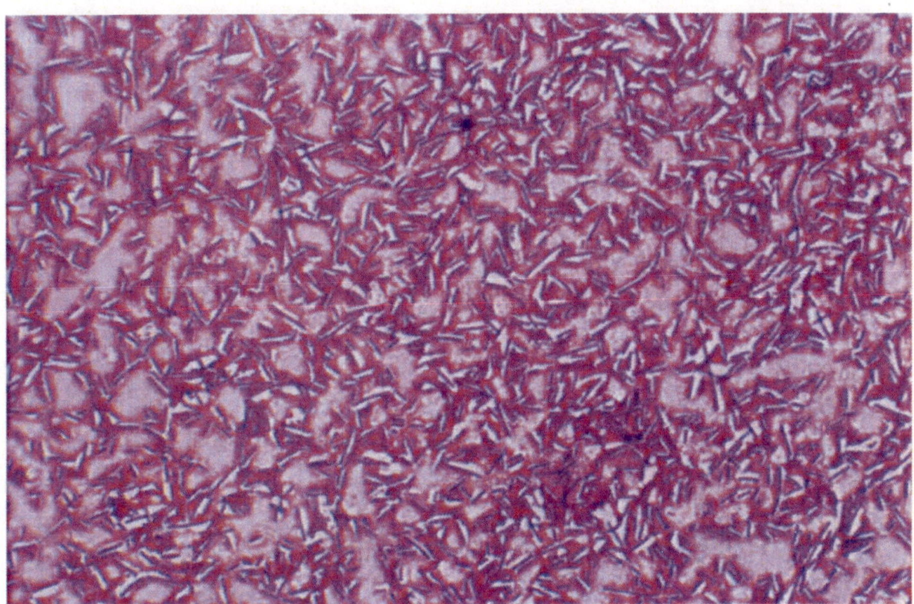

Mastitis – *Klebsiella organisms*-Gram negative non motile rod shaped bacteria Gram's stain 100x

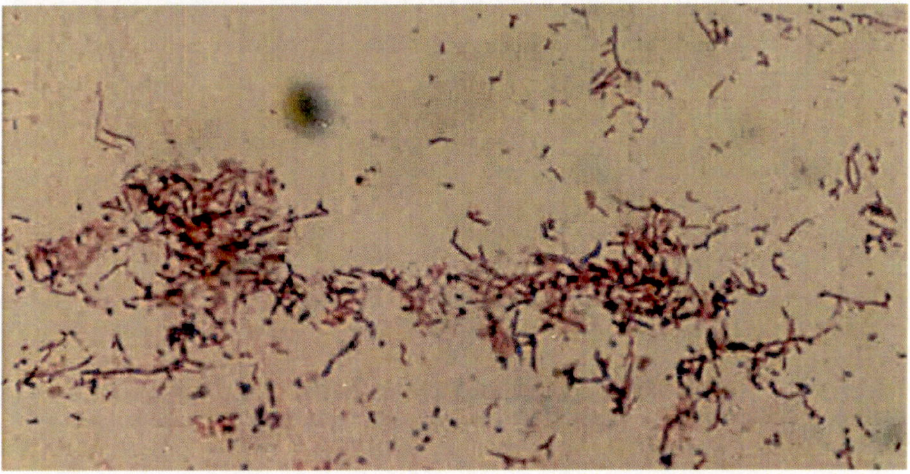

Nocardial mastitis- Presence of Gram positive bacillary branching bacteria Gram's stain 100x

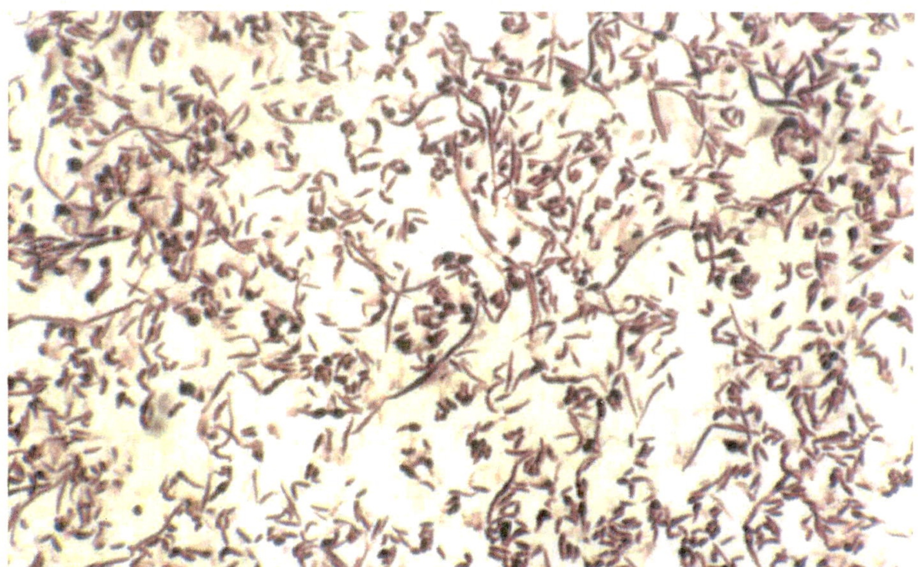

Actinomyces bovis mastitis- Presence of straight or slightly curved bacteria Leishman & Giemsa 100x

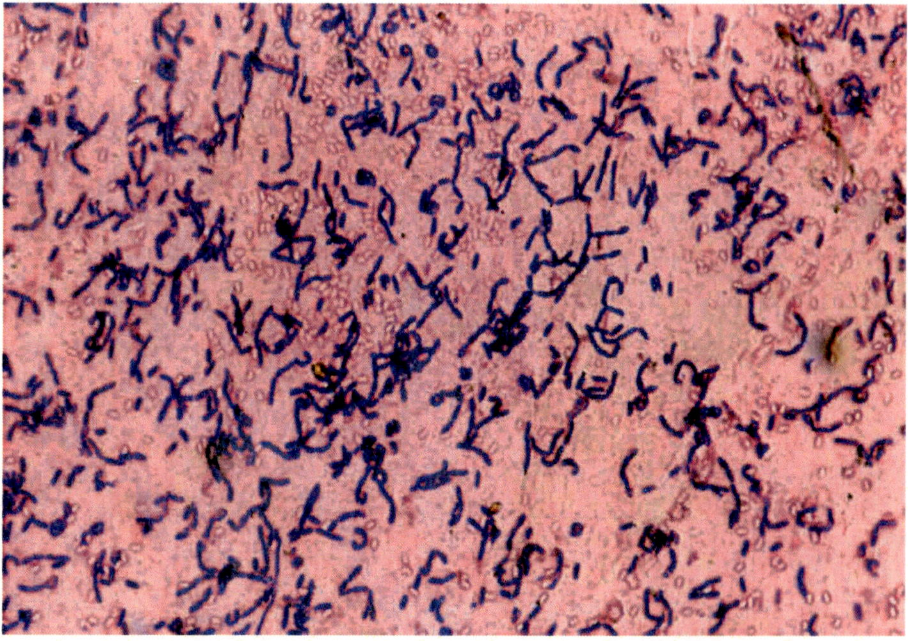

Actinomyces bovis mastitis- Presence of straight or slightly curved bacteria Gram's stain 100x

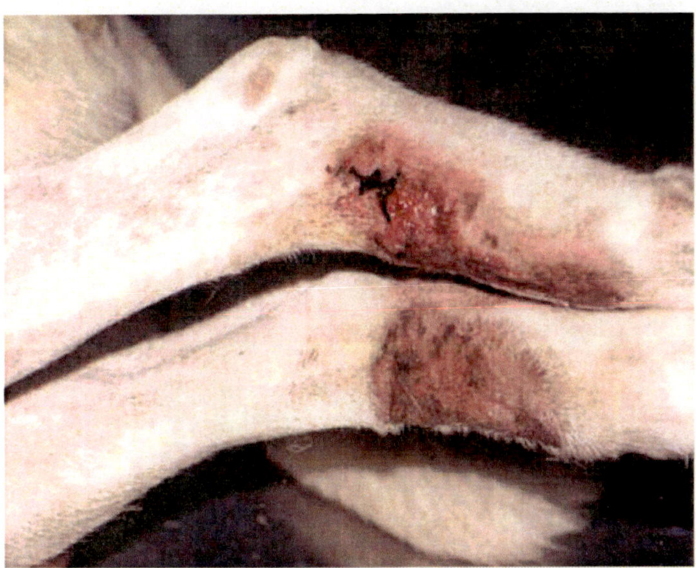

Acral lick dermatitis-Non-Descriptive Dog- Near the hock joint

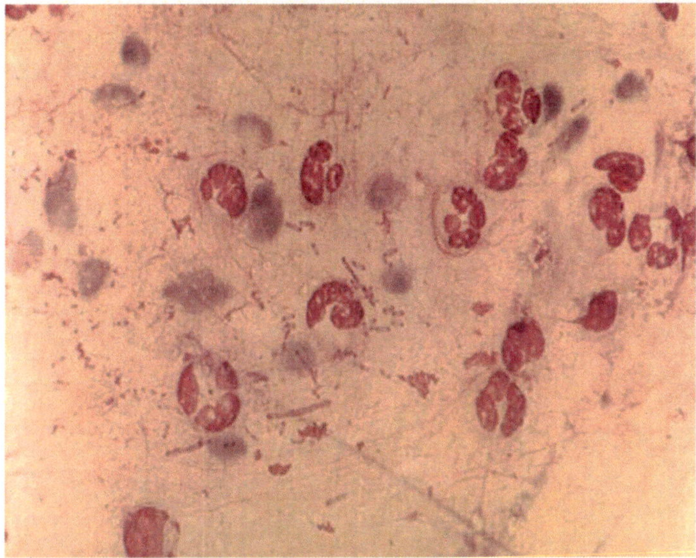

Acral lick dermatitis – Cytology - Presence of intact neutrophils and bipolars Leishman- Giemsa stain 100X

Parafilariasis

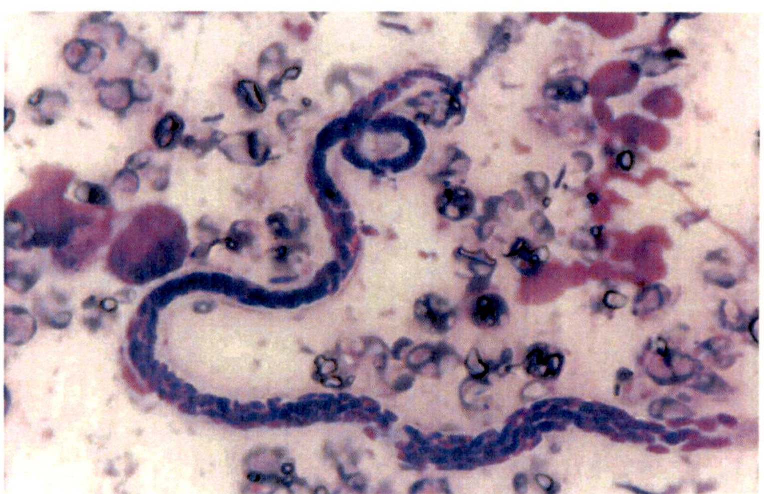

Cattle-Parafilariasis- *Parafilaria bovicola* Leishman & Giemsa 100x

Specific Conditions

Canine Distemper

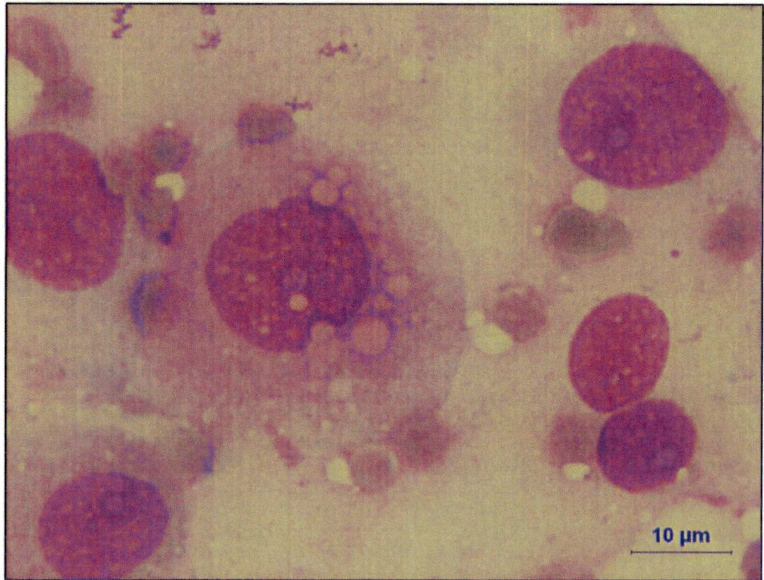

Canine distemper- Necropsy - Impression-Inclusions in the urinary bladder- Numerous inclusions bodies in the transitional epithelial cell cytoplasm Leishman & Giemsa 1000x

Theileriosis

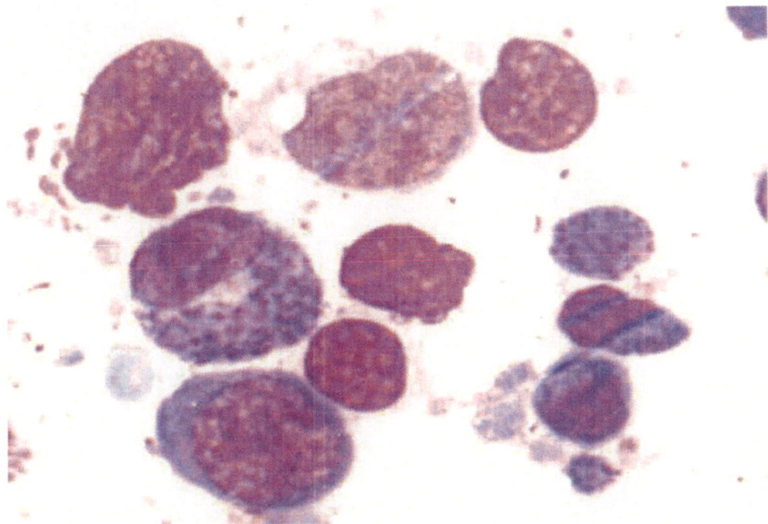

Theileriosis- Cattle-Lymph node aspirate-Presence of giant schizonts of *Theileria* spp. organisms known as Koch blue bodies in the lymphocytes Leishman & Giemsa 1000x

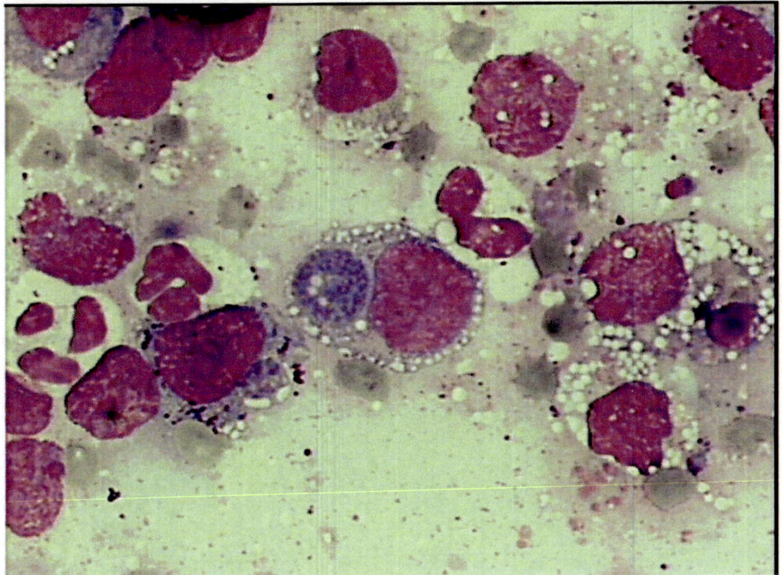

Theileriosis- Lymph node smear- Koch's blue bodies- Presence of giant schizonts of *Theileria spp.* organisms in the lymphocytes Leishman & Giemsa 1000x

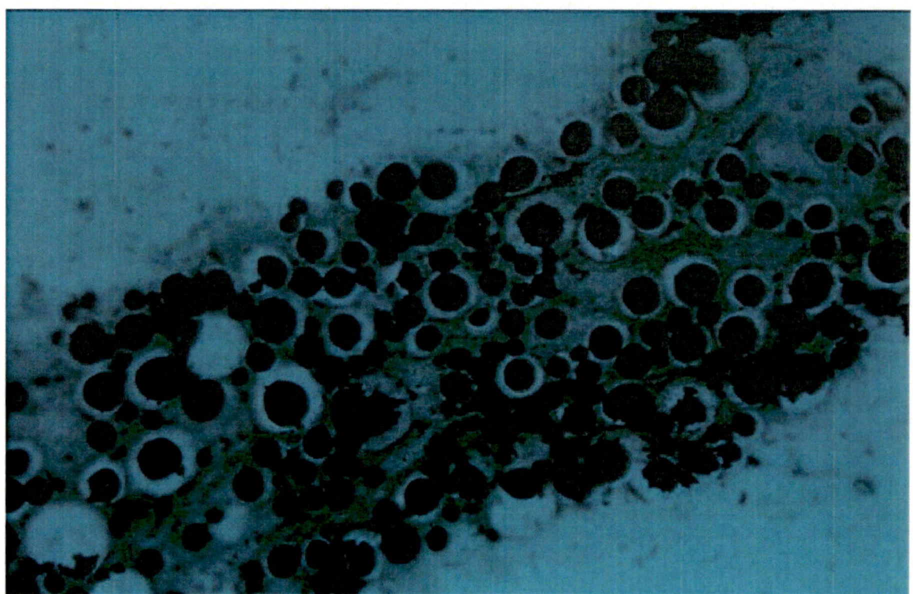

Dog – Blastomycosis -Thick cell walled with broad based budding yeast - Grocott-Gomori's Methenamine Silver (GMS STAIN) x 400

Fatty Degeneration

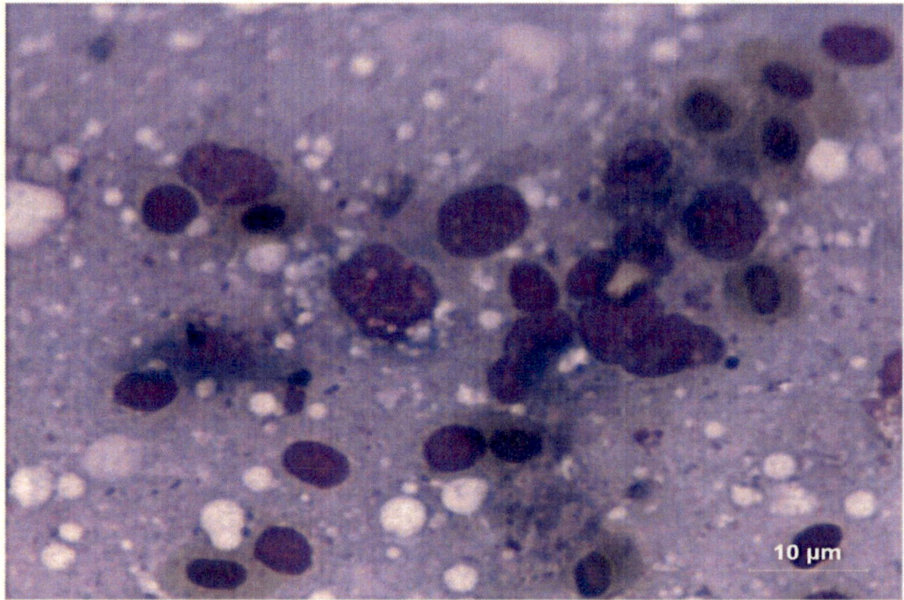

Liver- Fatty degeneration- Vacuoles in the cytoplasm of hepatocytes Leishman & Giemsa 10 µm

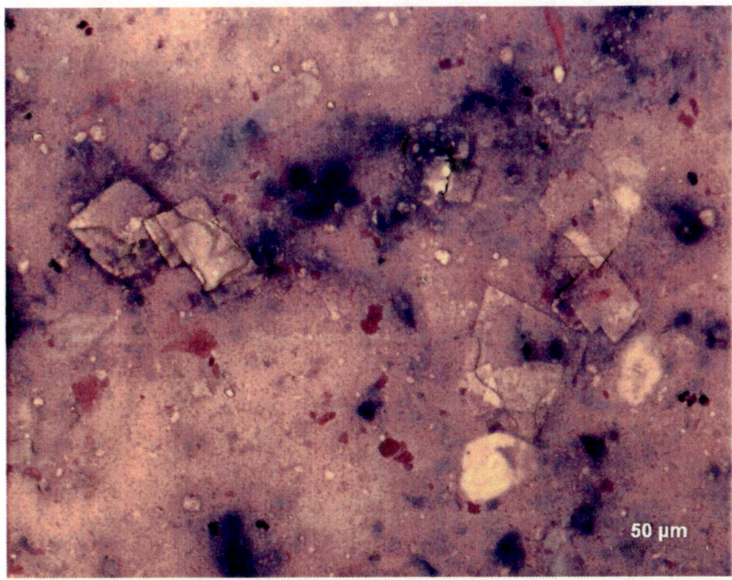

Dermoid cyst-cholesterol clefts-Dog Leishman & Giemsa Bar 50 µm

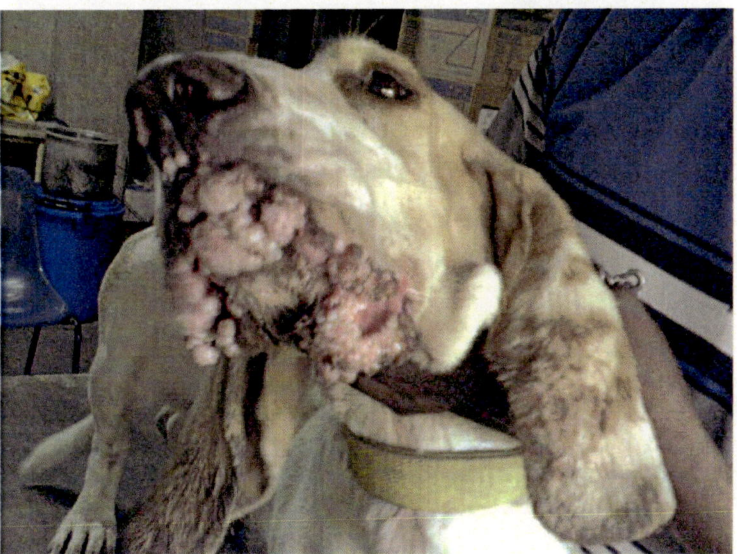

Dog-Oral papilloma/Dog/Oral

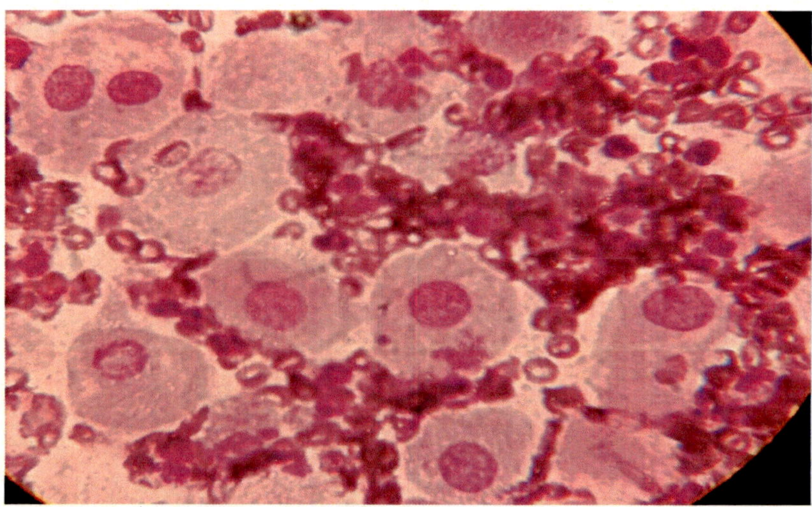

Papilloma -Dog-Skin- Round squamous cells. Emperipolesis Leishman & Giemsa x100

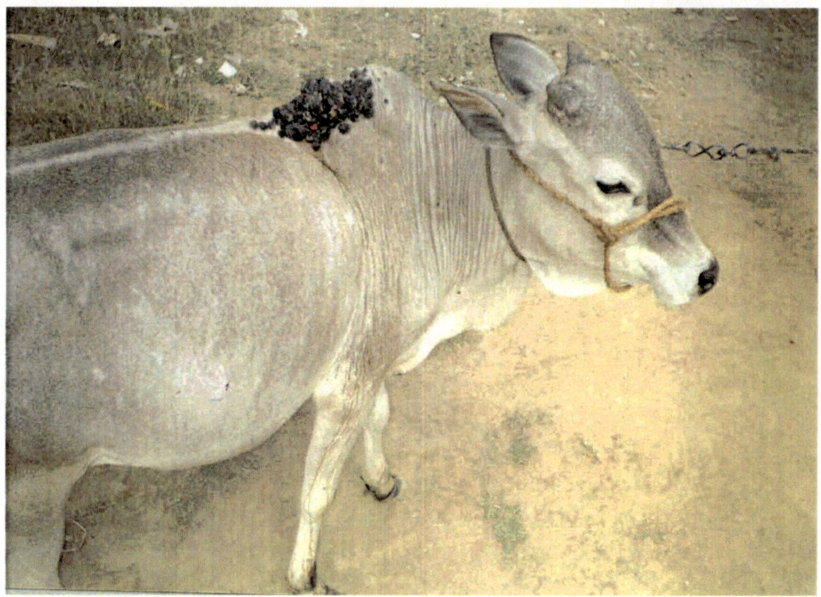

Papilloma –Wart- Calf

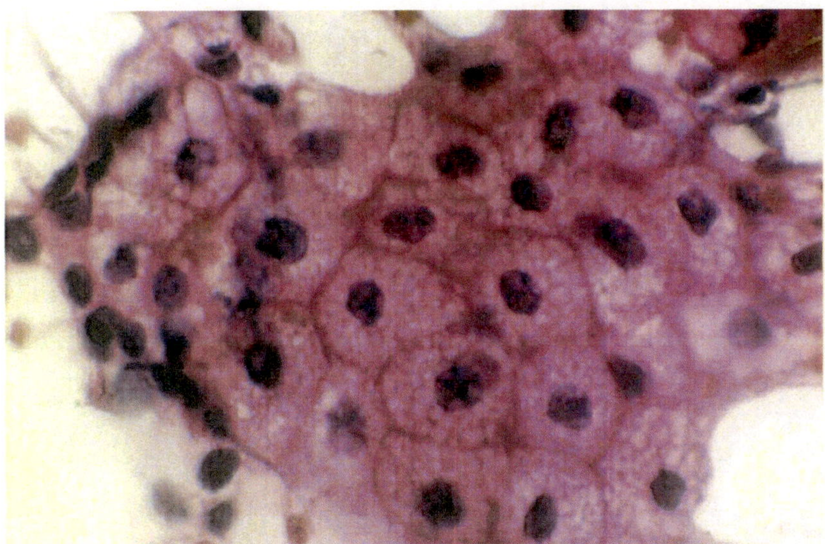

Papilloma-Dog-Cluster of squamous cells with round to oval nuclei Leishman & Giemsa x100 *(Courtesy:* Krithiga *et al.*, 2005)

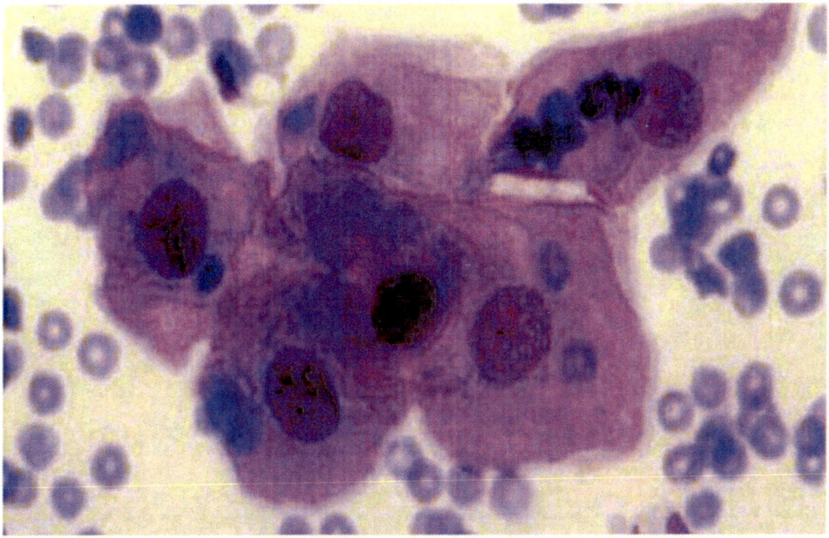

Papilloma-Dog- Cluster of angular squamous cells with round to oval nuclei Leishman & Giemsa x100 (*Courtesy:*Krithiga *et al.*, 2005)

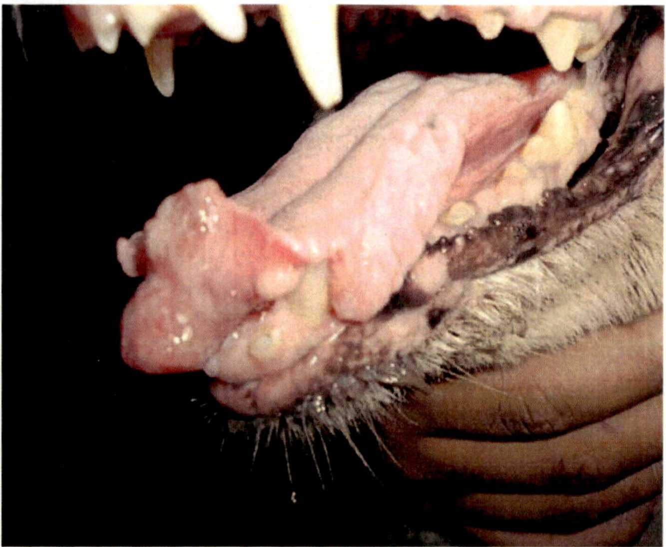

Squamous cell carcinoma- Spitz dog-Tongue –Spherical gray white colored firm mass

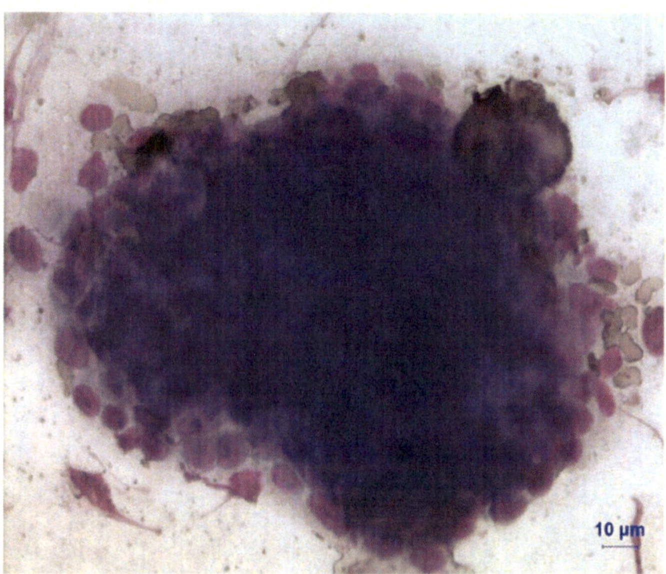

Squamous Cell Carcinoma- Clusters of neoplastic squamous cells with basophilic cytoplasm. Leishman & Giemsa Scale Bar 10 µm

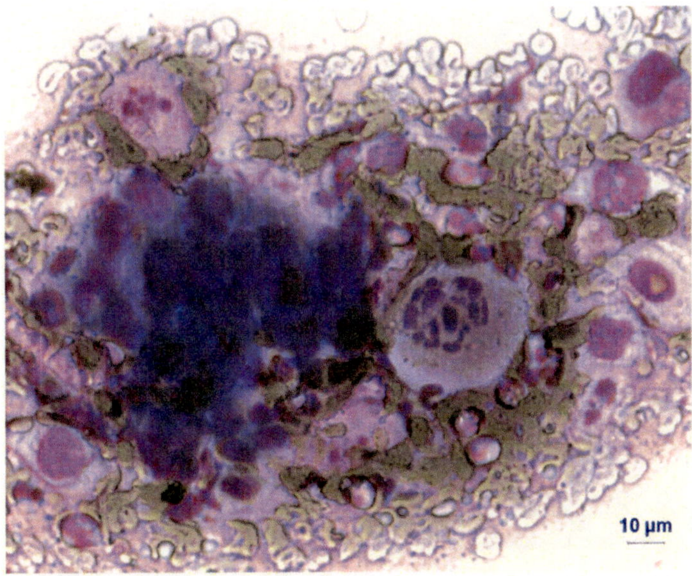

Squamous cell carcinoma- Cluster of cells- Mitotic figure. Leishman & Giemsa Scale Bar 10 µm

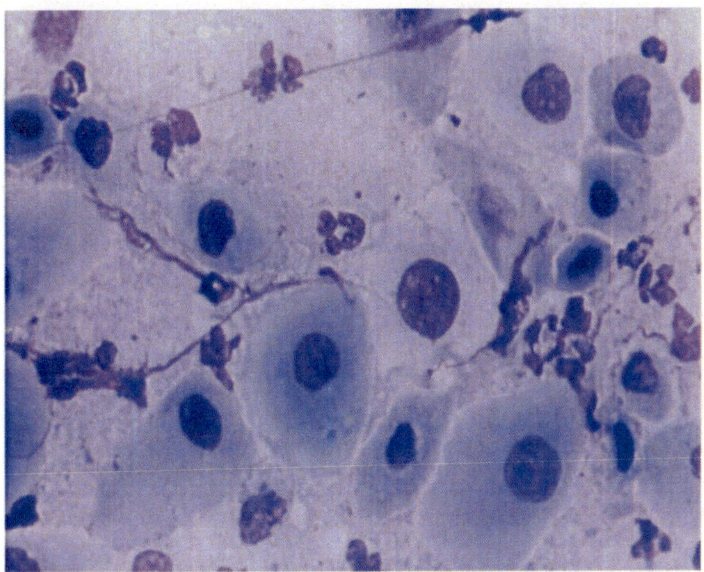

Squamous cell carcinoma- Cluster of squamous cells with round to oval nuclei. Basophilic cytoplasm. A few neutrophils are seen.

Leishman &Giemsa x100 (Courtesy: Krithiga *et al.*, 2005)

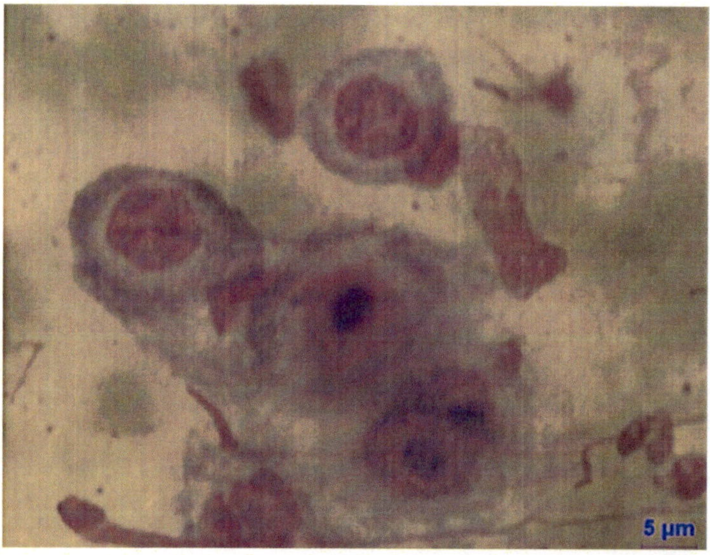

Squamous cell carcinoma- Large basophilic nucleoli with vacuolation of cytoplasm. Leishman & Giemsa Scale bar 5 µm

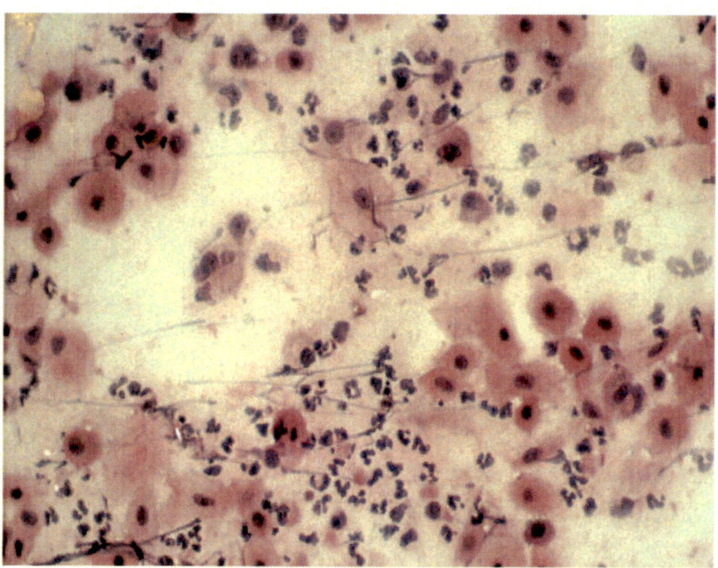

Squamous cell carcinoma- Dog-Inflammatory carcinoma- Cluster of squamous cells and numerous neutrophilic infiltration Haematoxylin & Eosin x100 (*Courtesy*: Krithiga *et al*., 2005)

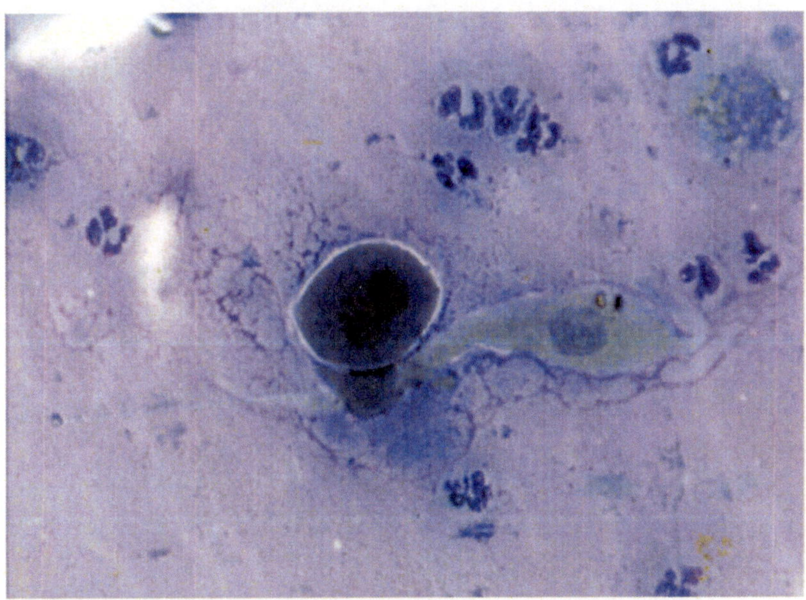

Squamous cell carcinoma- Tadpole cell indicates squamous cell carcinoma. Leishman & Giemsa x100. (Courtesy: Krithiga *et al.*, 2005)

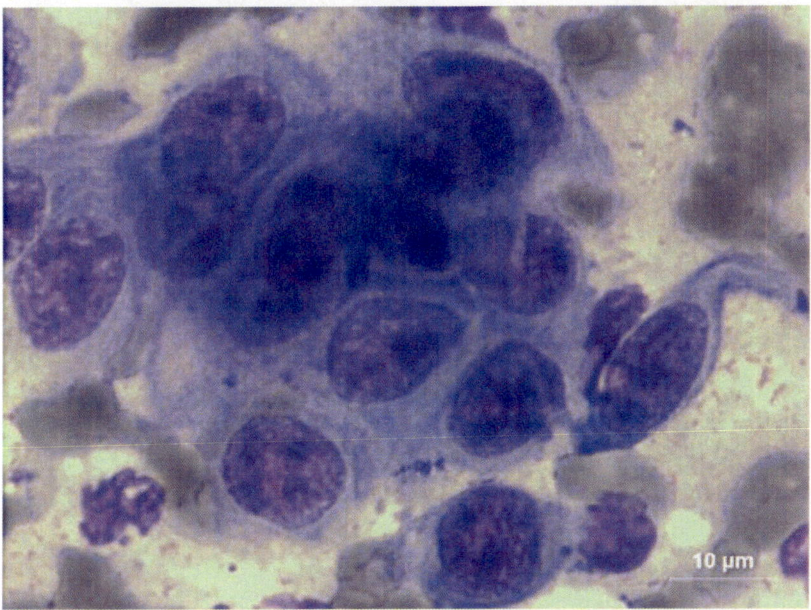

Squamous cell carcinoma- Tadpole cell. Leishman & Giemsa Scale bar 5 µm

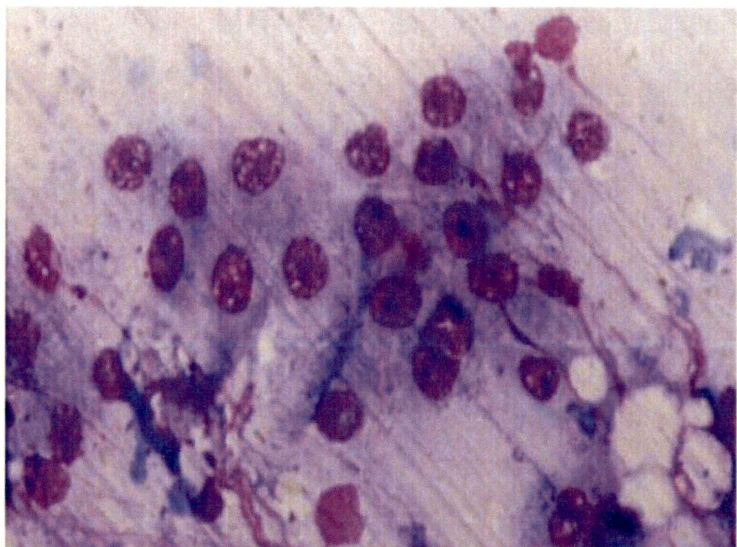

Sebaceous gland adenocarcinoma - Cluster of neoplastic sebocytes with round to spherical nuclei and basophilic cytoplasm with vacuolations. Leishman & Giemsa x400

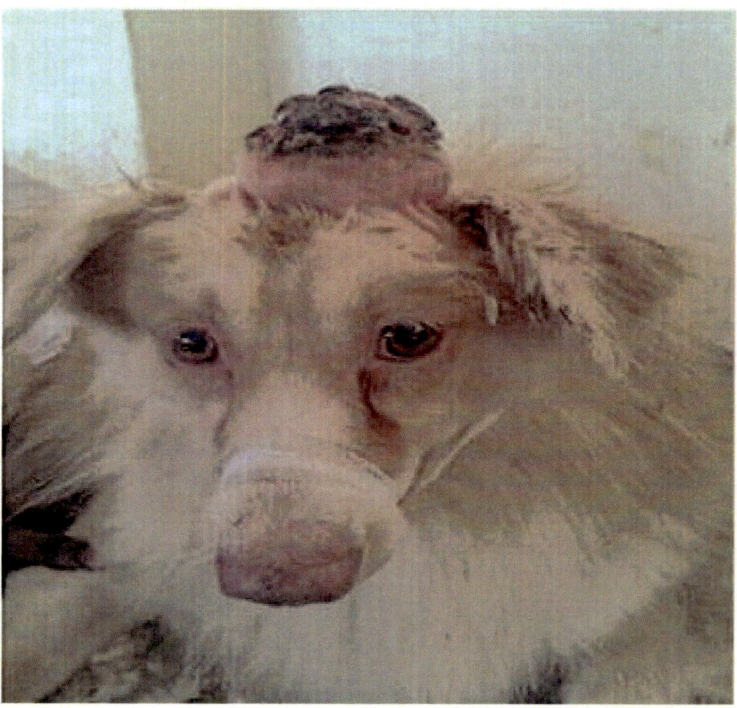

Trichoblastoma – Spitz- Forehead- Well circumscribed pink colored mass

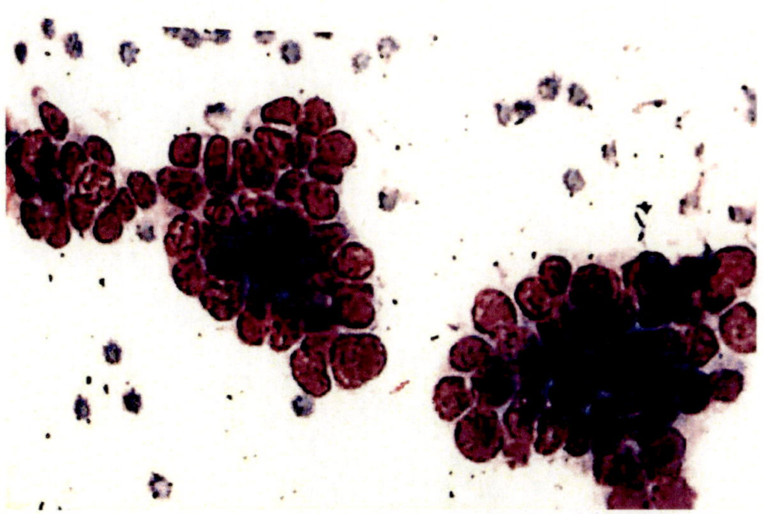

Basal cell carcinoma –Clusters of tightly adherent cylindrical cells with oval nuclei and sparse cytoplasm. Wright's Giemsa stain X 800 (Courtesy: Krithiga *et al.*, 2005)

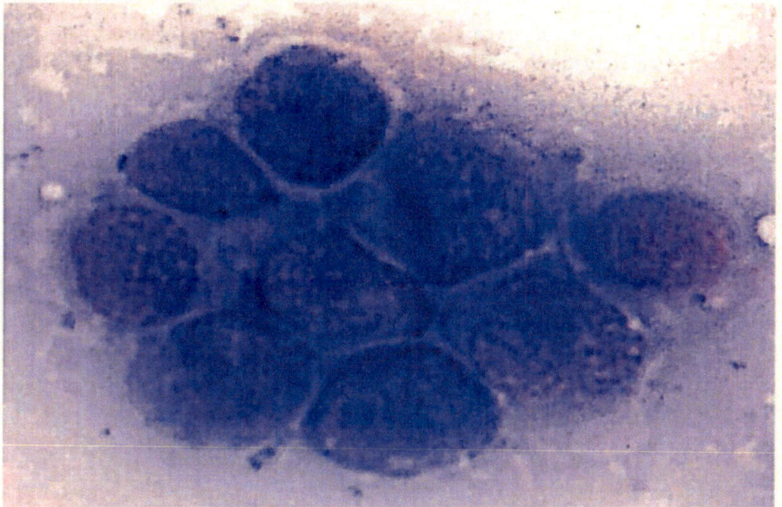

Perianal gland adenoma- Hepatoid cell cluster with spherical nucleus and granular chromatin. Wright's stain x 1000 (Courtesy: Krithiga et al., 2005)

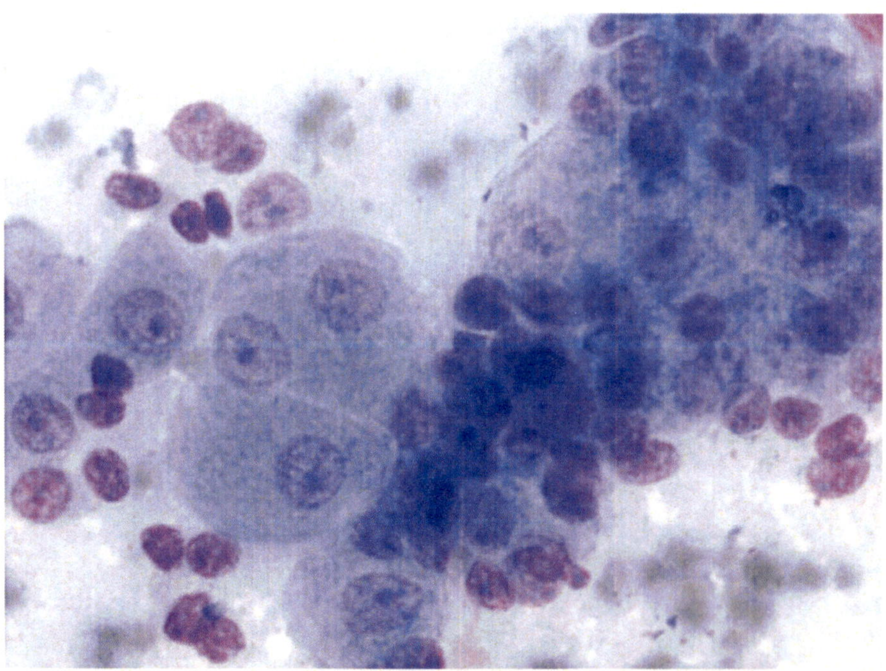

Perianal gland adenocarcinoma -Clusters of pleomorphic hyperchromatic hepatoid cells with basophilic cytoplasm. Binucleated cell and basal cell also seen. Leishman & Giemsa x1000. (Courtesy: Krithiga *et al.*, 2005)

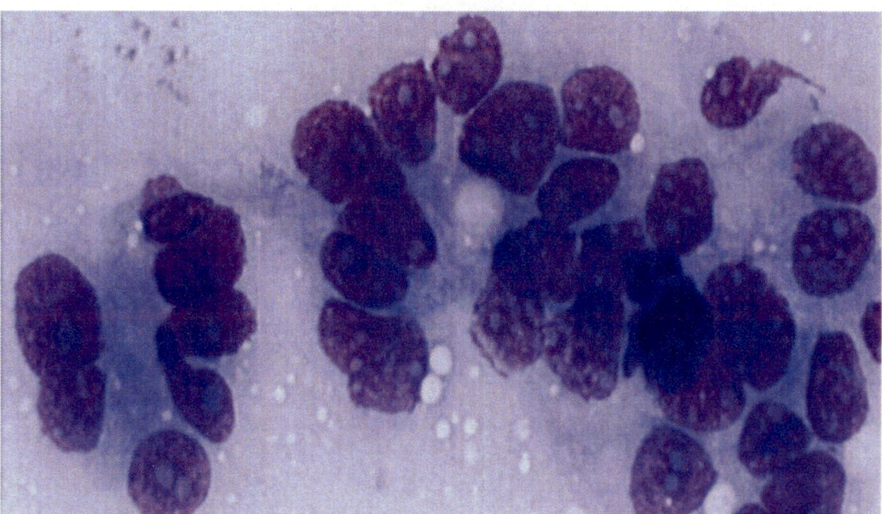

Ethmoid tumour- Adenocarcinoma- Epithelial cells arranged in acinar pattern with prominent multiple nucleoli in the vesicular nuclei and coarse chromatin. Basophilic cytoplasm was seen. Leishman & Giemsa x1000

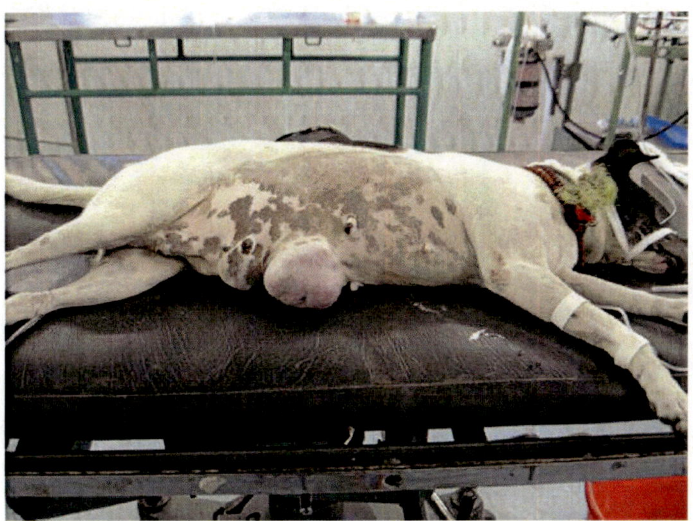

Mammary tumour-Dog-Primary tumour

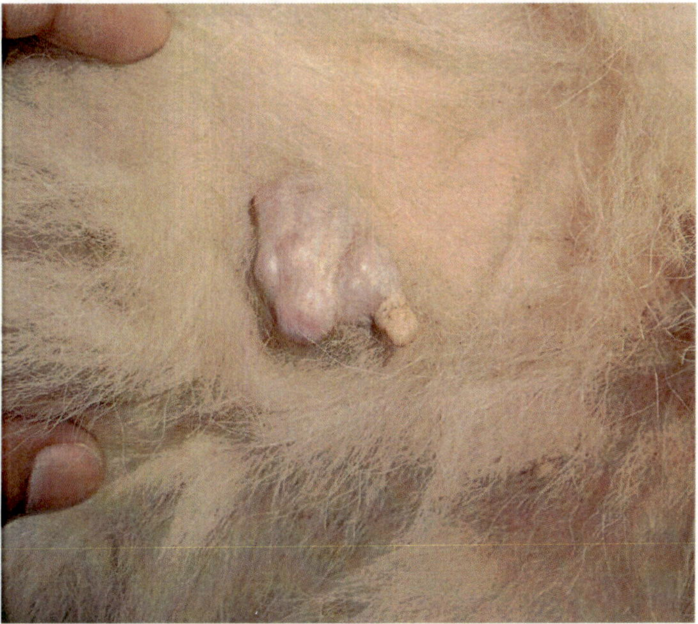

Mammary adenocarcinoma – Spitz – Female -Multilobulated mass

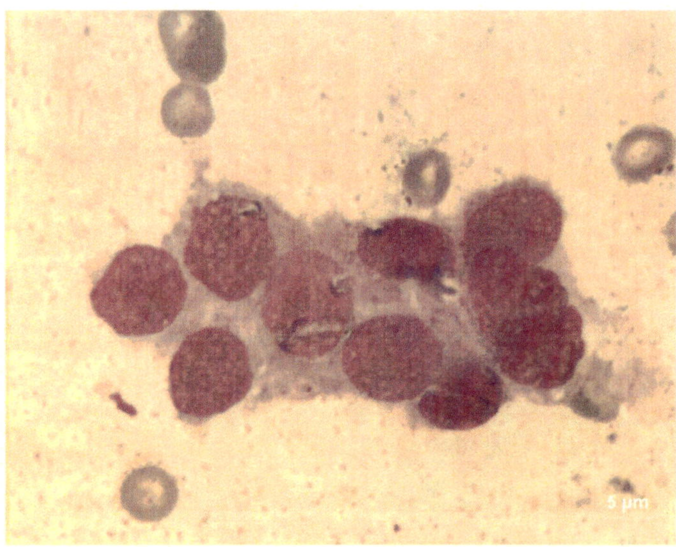

NAF- Canine mammary tumour-Benign cells- Cuboidal uniform sized cells with round to nuclei and moderate light pale basophilic cytoplasm. Acinar pattern. Leishman & Giemsa x 100

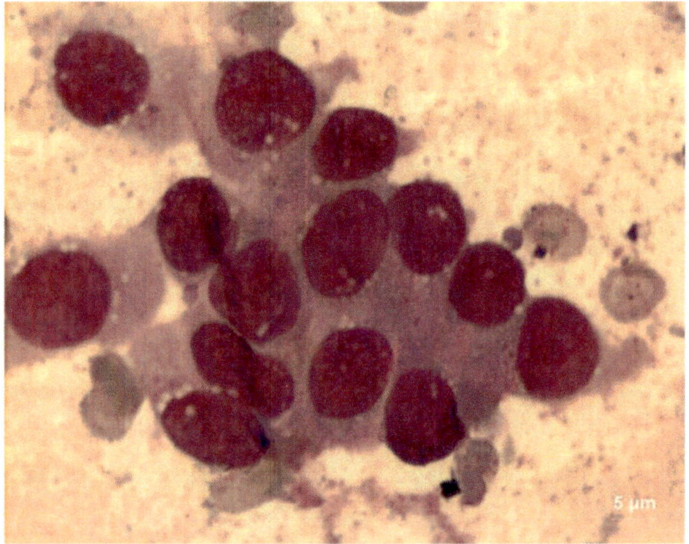

NAF- Canine mammary tumour-Benign cell clusters- Cuboidal cells with round to ovoid nuclei and moderate light pale basophilic cytoplasm in acinar arrangement. Leishman & Giemsa x 100

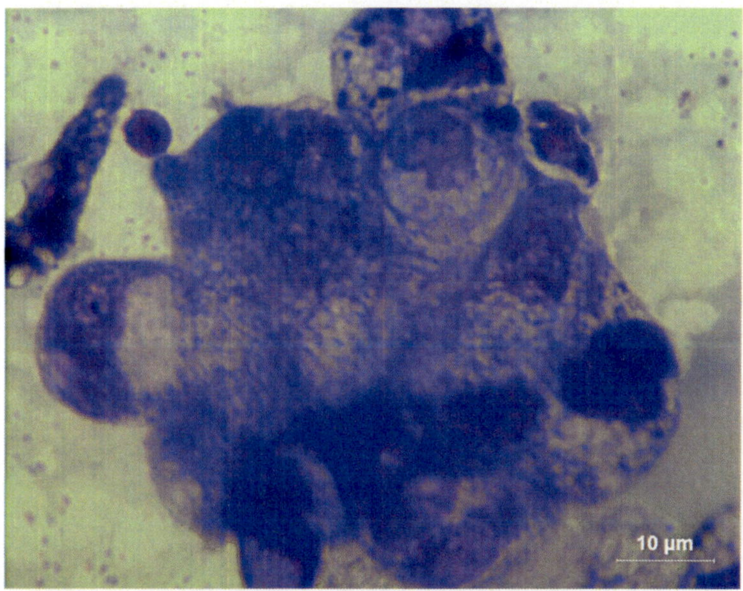

NAF- Adenoma- Cluster. Leishman & Giemsa Scale bar 10 µm

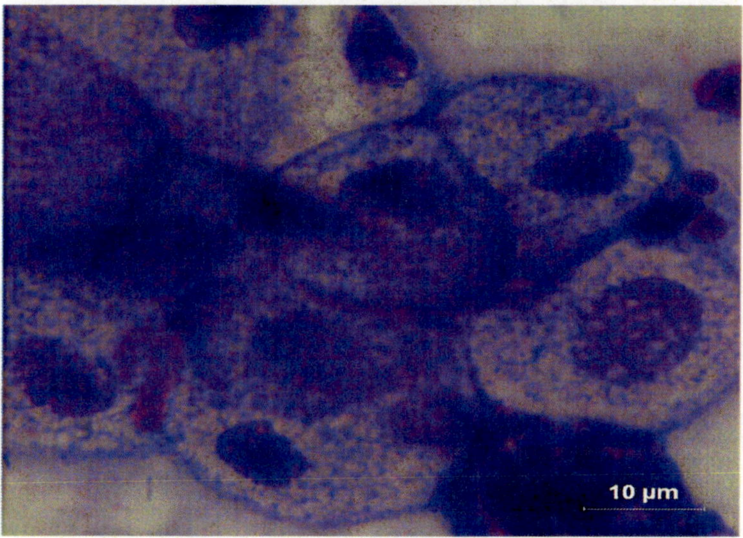

NAF-Carcinoma-Cluster of cells- Anisocytosis and anisokaryosis Leishman & Giemsa Scale bar 10 µm

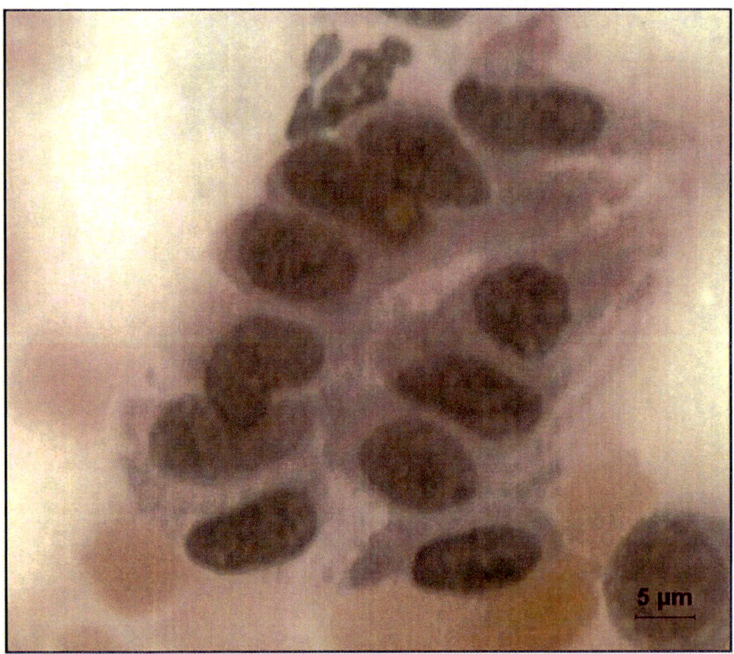

NAF- -Malignant variable sized cells in acinar pattern. Nucleus-Round to oval/cylindrical. Nucleoli present. Leishman & Giemsa Scale bar 5 µm

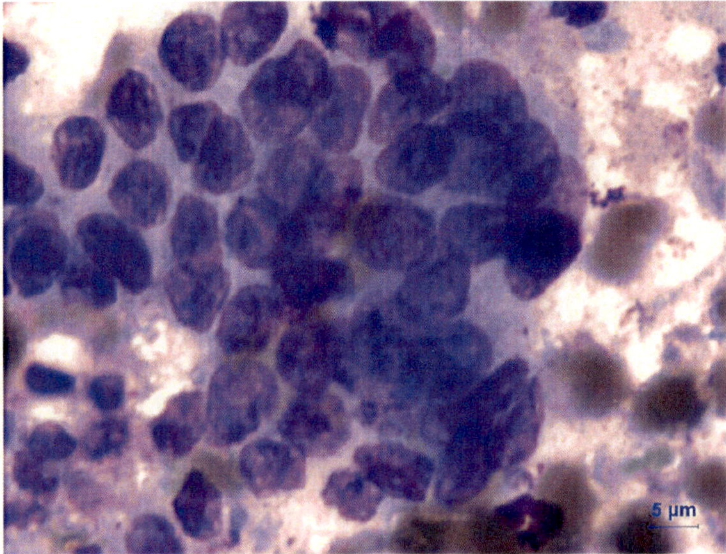

Mammary adenocarcinoma –Cluster of spherical to oval cells with round to oval nuclei and prominent nucleoli. Leishman & Giemsa Scale Bar 5 µm

12

Cytological Diagnosis of Mesenchymal Cell Tumours

Mesenchymal Tumours

General Features

- Do not exfoliate many cells – hypocellular; Low cellularity
- Embedded in extracellular matrix-osteoid, chondroid or collagen
- Single, discrete cells and a few are also seen in dense clumps i.e. Individual or unorganized cluster of spindle cells
- Cells-Elongated, fusiform/spindle shaped
- Cytoplasmic extensions/tails seen on both ends of nucleus; borders ill-defined
- Nucleus elongated/fusiform to oval
- As the malignancy potential increases cells become plumpy with a larger, rounder nucleus and shorter cytoplasmic tails
- Nuclei may be fusiform or oval.
- These tumors cannot be differentiated cytologically.
- Tentative diagnosis made.
- Well-differentiated malignant mesenchymal tumors may resemble actively proliferating granulation and requires histologic confirmation.

Fibromas

- Yield a few cells on aspirate or imprint; i.e. Exfoliate poorly
- Scraping may yield more cells depending on the type and matrix present
- Individual or group of cells seen, or in small aggregates
- Collagen background: Amorphous pink staining

- Cells uniform in size and shape, scant to moderate light blue cytoplasm in spindle, delicate bipolar tendrils and wisps
- Poor cell borders
- Aniso-cytosis/karyosis mild
- Nuclei: Round or oval/elongated stain with medium to marked intensity; Lacy chromatin
- One or two small, round indistinct basophilic nucleoli
- N/C ratios high
- Cytological features indistinguishable from fibroplasia

Fibrosarcoma

- Yield more cells than fibroma i.e. Exfoliate variably well, individually or in small aggregates
- Background: Bright-pink extracellular matrix
- Cells have variable volumes of cytoplasm forming bipolar tendrils and wisps
- Cells less spindle, plumpy or round shaped; some stellate or have one tail
- May contain fine pink granules or clear vacuoles
- Nuclei ovoid to elongated, finely granular chromatin
- Prominent nucleoli
- Aniso-cytosis/karyosis often marked
- Mitosis and occasional multinucleation may be present.
- As malignancy increases cytoplasmic basophilia, high N:Cy ratio, aniso-cytosis/karyosis/ nucleoliosis are observed

Granulation Tissue

- Fibrosarcoma may resemble granulation tissue.
- Granulation tissue comprises proliferating fibroblasts, young plump and capillaries, difficult to differentiate from anaplastic neoplasm. Histopathology can only confirm the case.

Chondrosarcoma

- Large amount of matrix; Dense extracellular chodroid-characteristic, in which cells will be embedded forming lacunae
- Cells: Round or spindloid having pale-blue cytoplasm containing diffuse, fine-pink granules.
- Unlike osteosarcomas, nuclei are centrally placed with finely stippled chromatin and multiple basophilic nucleoli

Osteosarcoma

- Highly cellular; Exfoliate well
- Associated with bright-pink, fibrillar ECM (Osteoid)
- Cells distributed individually and in aggregates which range from ovoid to fusiform
- Cytoplasm contains fine clear vacuoles or fine pink granules
- Round to ovoid eccentrically placed nuclei with appearance of falling out of cell
- Finely granular chromatin with multiple basophilic nucleoli, hyperchromasia
- Marked criteria of malignancy often are present incl. multinucleated cells
- Difficult to differentiate subtypes chondroplastic and fibroplastic from chondrosarcoma and fibrosarcoma

Neoplastic cells may resemble plasma cells with eccentric nuclei, abundant cytoplasm but lacking paranuclear clearing as seen in plasma cells referred to as "plasmacytoid cells".

Lipomas

- Occur commonly in the subcutaneous tissues of dogs and cats
- Lipomas appear wet/greasy/oily and smears do not dry completely on microscope slides.
- Cells often rupture
- Fat dissolves in alcohol containing fixative used in some stains, so the slide often appears acellular adipocytes cytologically.

- Adipocytes are very large cells which are distended with fat.
- Abundant clear cytoplasm seen
- Nuclei is eccentric; compressed by fat vacuole to a side

Liposarcomas

- Rare in occurrence
- Yield numerous cells; Exfoliate well
- Have abundant cytoplasm.
- Cell membrane may be difficult to appreciate cytologically.
- Cells that contain cytoplasmic vacuoles which vary in size and number.
- Fat vacuoles are seen in the background.
- Multinucleated cells may be present

Note: Lipoma and liposarcoma, aqueous based stains can be used to stain fat e.g. New Methylene Blue, Oil Red O, Sudan Red

When large cells distended with fat are seen cytologically, examine lipomatous tumor tissues histopathologically, as many malignant tumors invade subcutaneous fat and neoplastic cells can be detected cytologically.

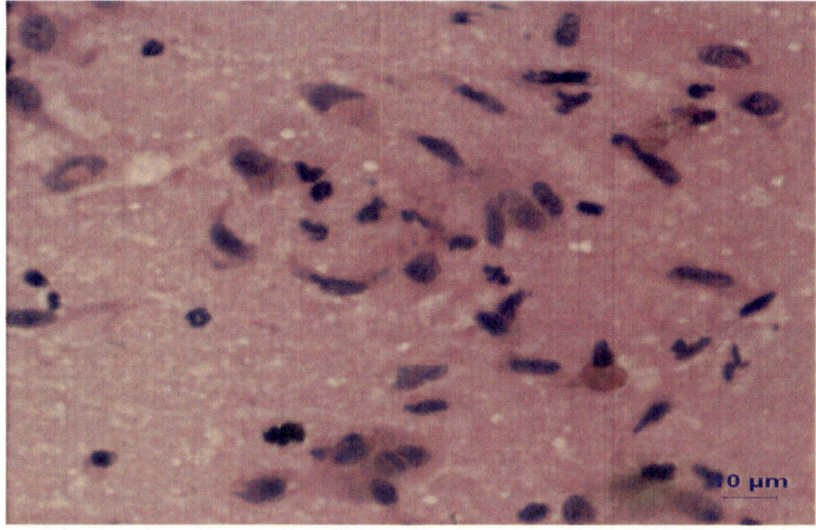

Fibroma- Spindle shaped cells with elongated to oval nuclei with indistinct light blue cytoplasm Leishman & Giemsa Scale Bar 10 µm

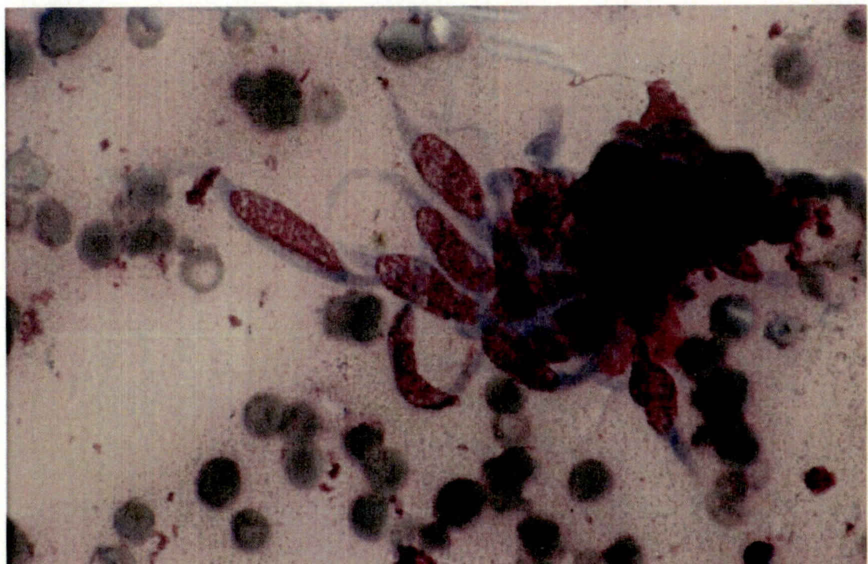

Fibroma- Uniform sized spindle shaped cells with elongated to oval nuclei with indistinct light blue cytoplasm Leishman & Giemsa x 100 (*Courtesy:* Krithiga *et al.*, 2005)

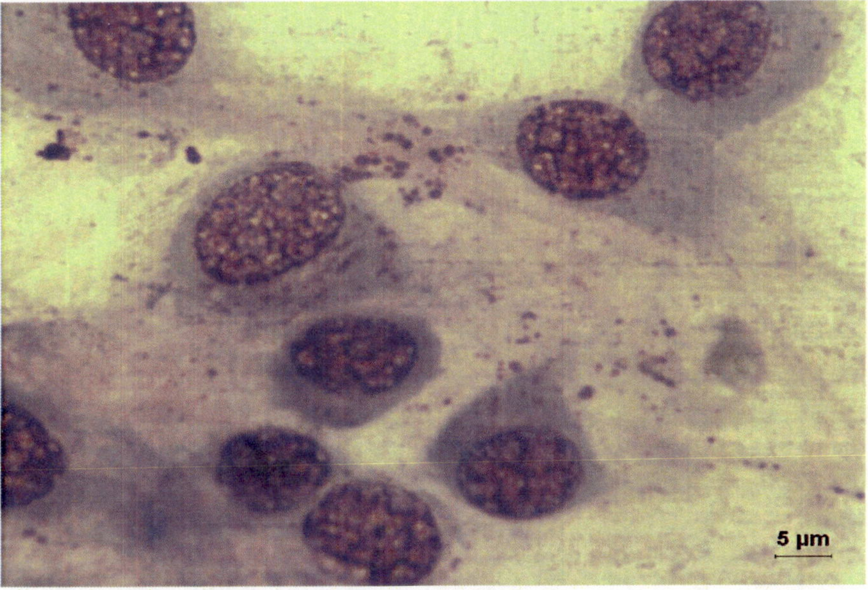

Fibrosarcoma-Partially spindle to plumpy cells with basophilic cytoplasm. Round to oval shaped nuclei with coarse chromatin. Prominent and multiple nucleoli Leishman & Giemsa x 5µm

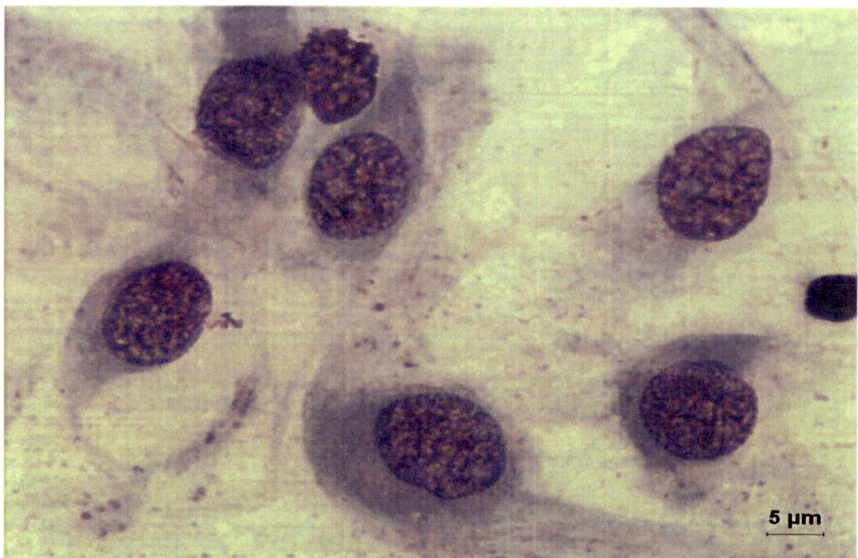

Fibrosarcoma- Unilateral or mild bilateral spindle to plumpy cells with basophilic cytoplasm. Round to oval shaped nuclei with coarse chromatin. Prominent and multiple nucleoli Leishman & Giemsa x 5µm

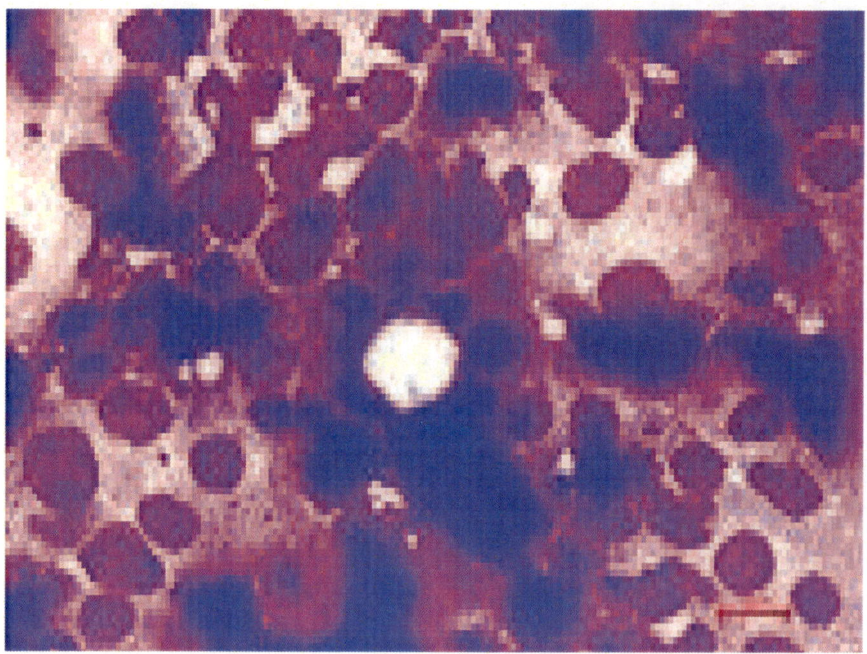

Haemangiosarcoma: Blood vessel lined by plumpy endothelial cells Leishman & Giemsa x 10µm

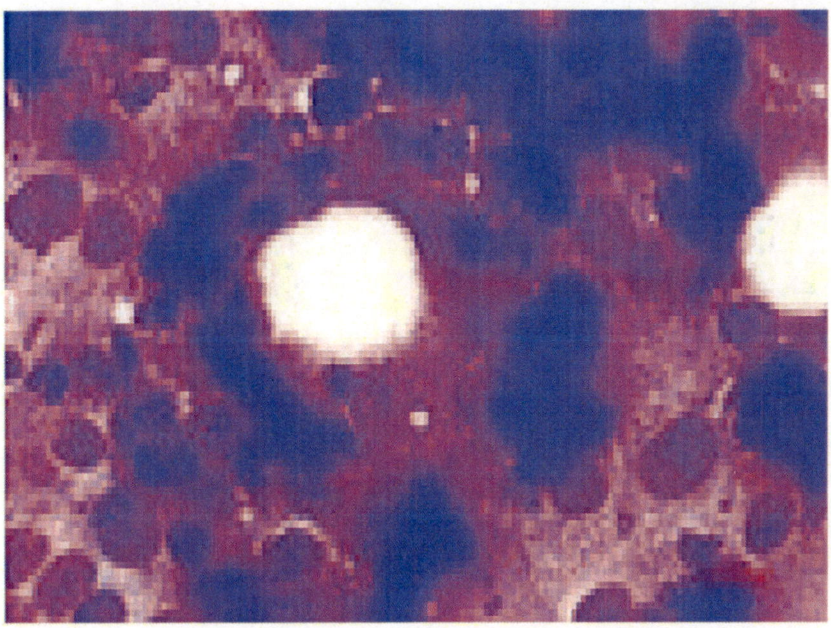

Haemangiosarcoma: Blood vessel lined by plumpy endothelial cells.
Leishman & Giemsa x 10μm

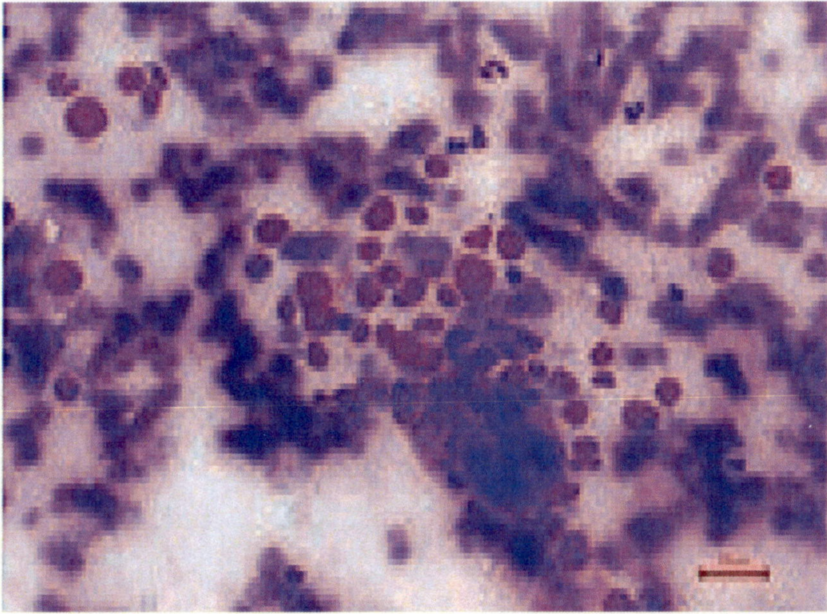

Haemangiosarcoma: Cluster of plumpy endotheilal cells Leishman & Giemsa x 10μm

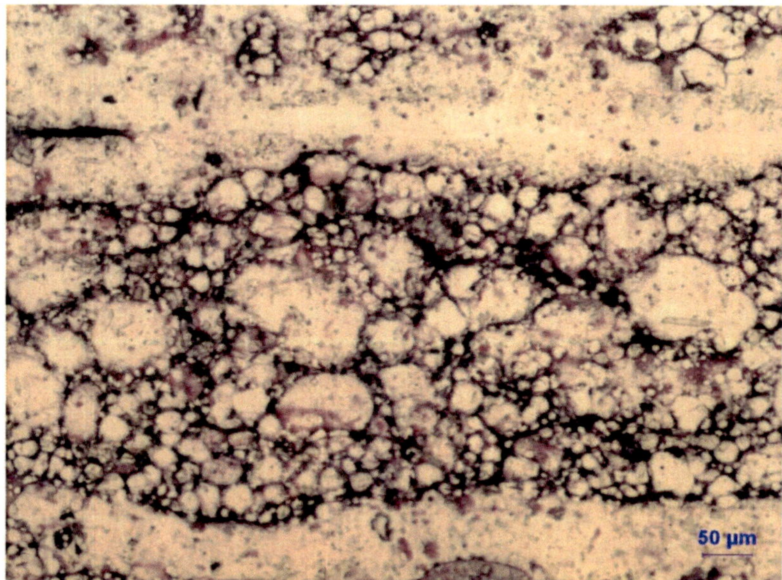

Lipoma-Variable sized neoplastic adipocytes with eccentric nuclei Leishman & Giemsa x 50µm (*Coutesy:* Krithiga *et al.*, 2005)

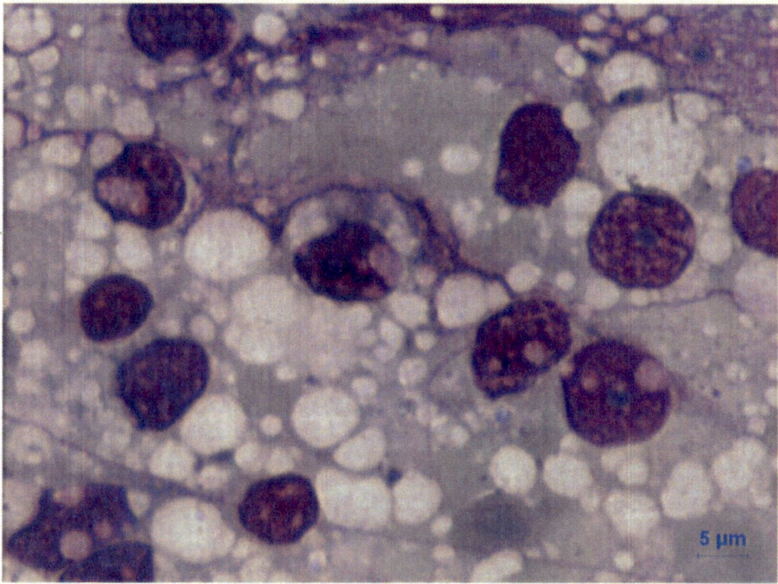

Liposarcoma-Round to ovoid cells with large round nuclei and varied amount of pale cytoplasm. Prominent nucleoli are seen Leishman & Giemsa x 5µm

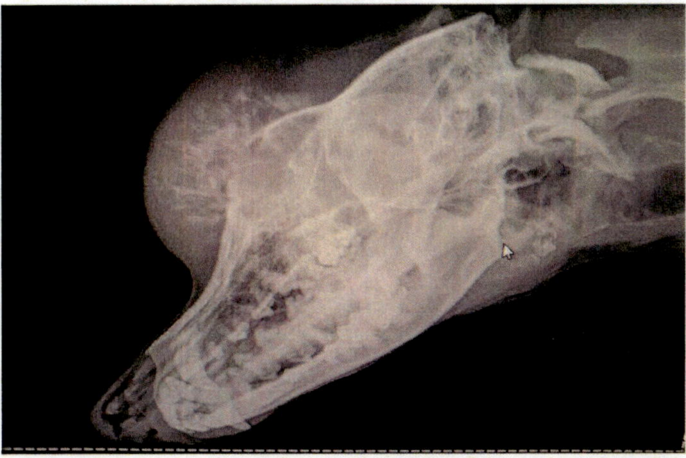

Skull- Radiography-Osteosarcoma- Sun burst appearance

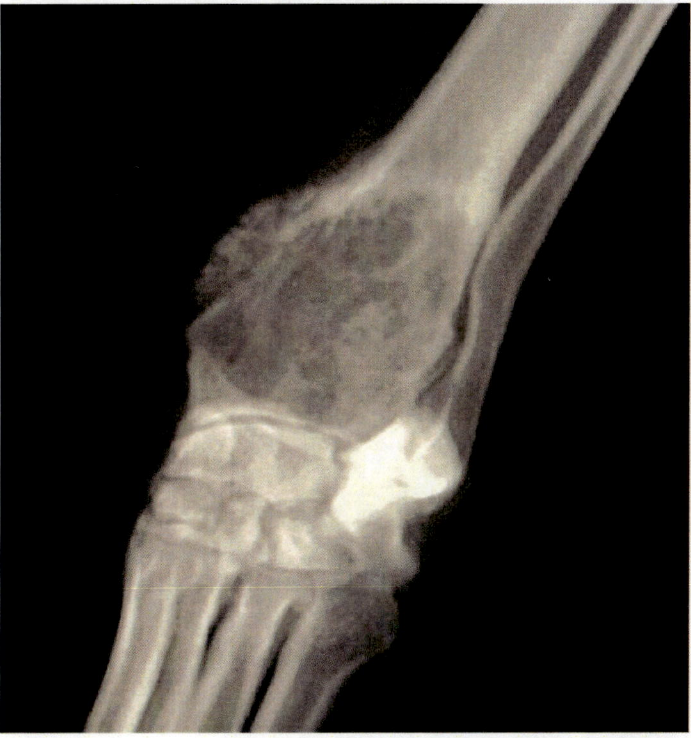

Radius and ulna- Radiography-Osteosarcoma- Sun burst appearance

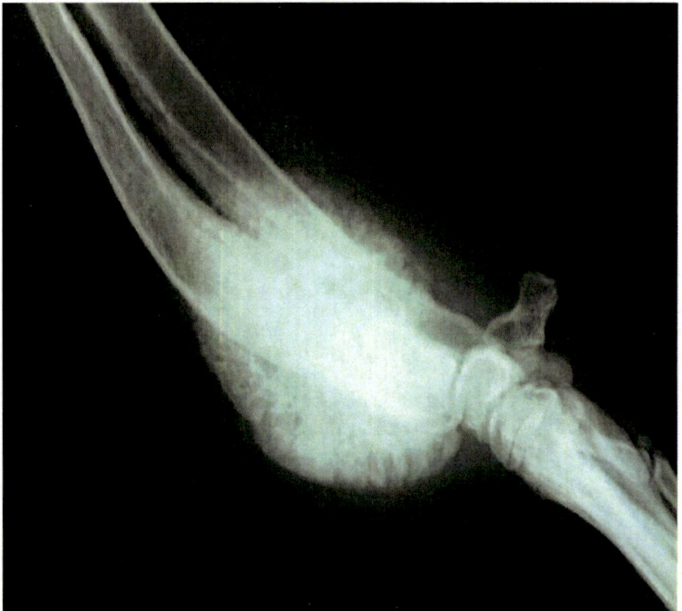

Radius and ulna- Radiography- Osteosarcoma – Sun burst appearance

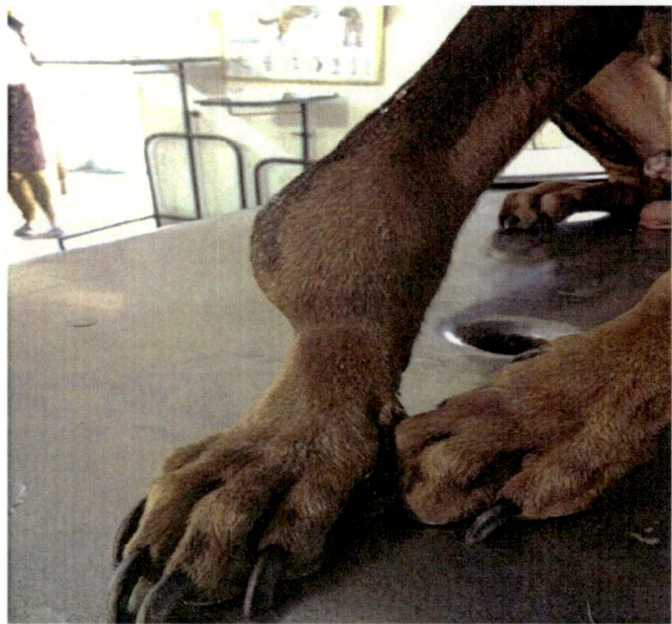

Distal part of radius and ulna - Osteosarcoma – Doberman
– Round hard growth

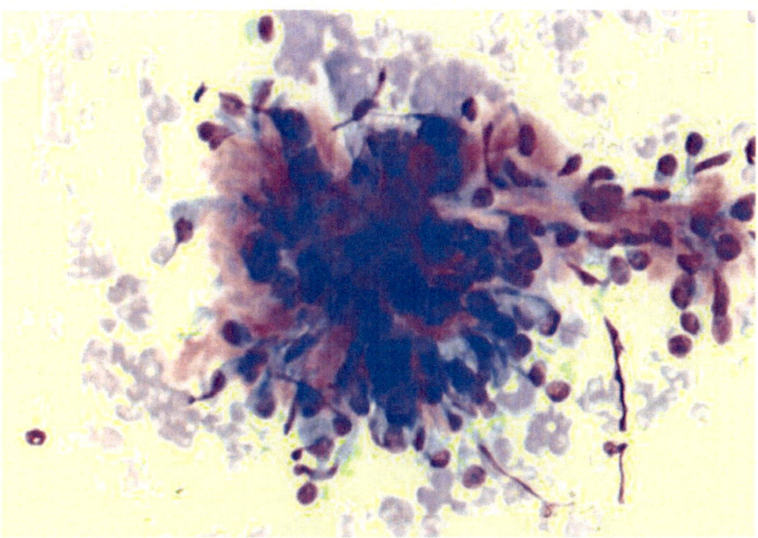

Osteosarcoma- Numerous spindled mesenchymal cells embedded in eosinophilic matrix. Leishman & Giemsa 100x

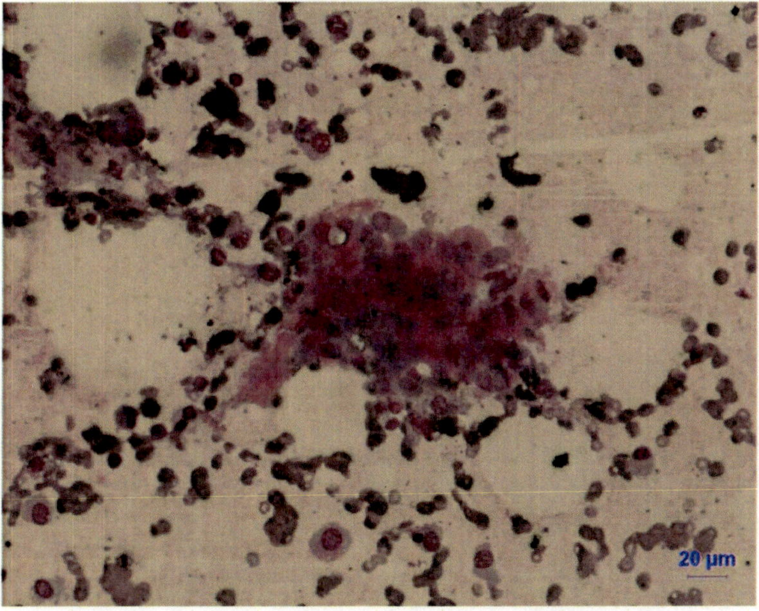

Osteosarcoma – Cluster of osteoblastic cells and pink stained matrix. Leishman & Giemsa Scale Bar 20µm

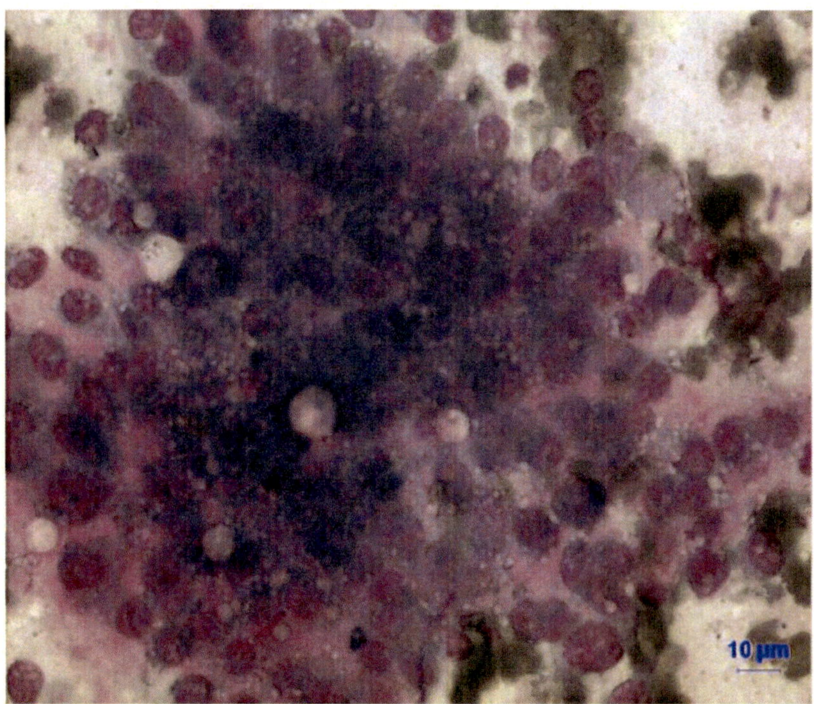

Osteosarcoma –Round to oval cells with fine cytoplasmic vacuoles. Leishman & Giemsa Scale Bar 10 µm

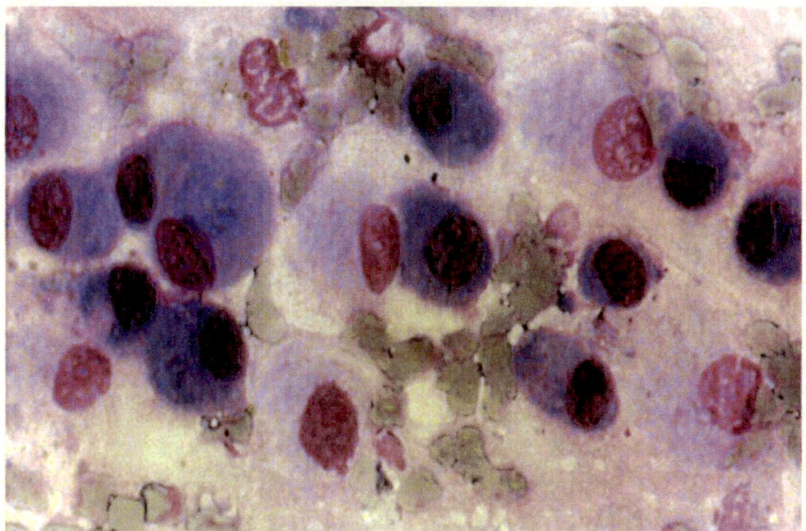

Canine Osteosarcoma-Plasmacytoid cells and abundant basophilic cytoplasm. Leishman & Giemsa 100x (*Courtesy:* Krithiga *et al.*, 2005)

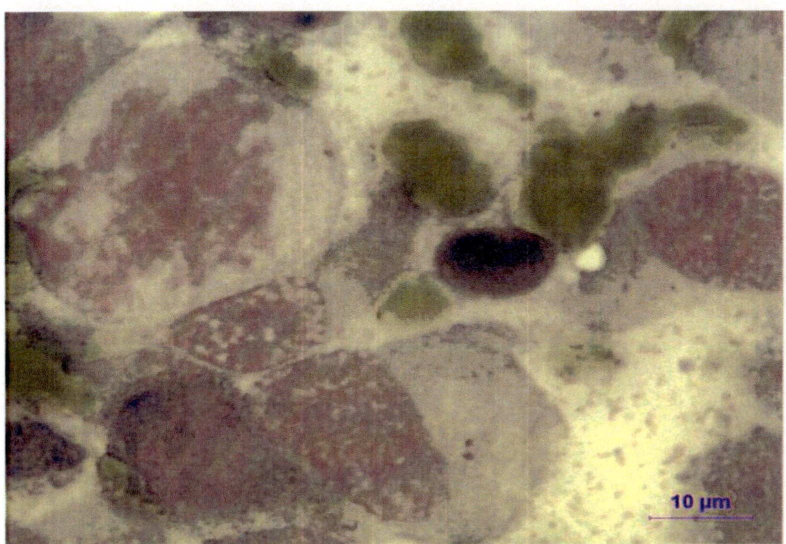

Osteosarcoma- Plasmacytoid cells with mitotic figure. Leishman & Giemsa Scale Bar 10 µm

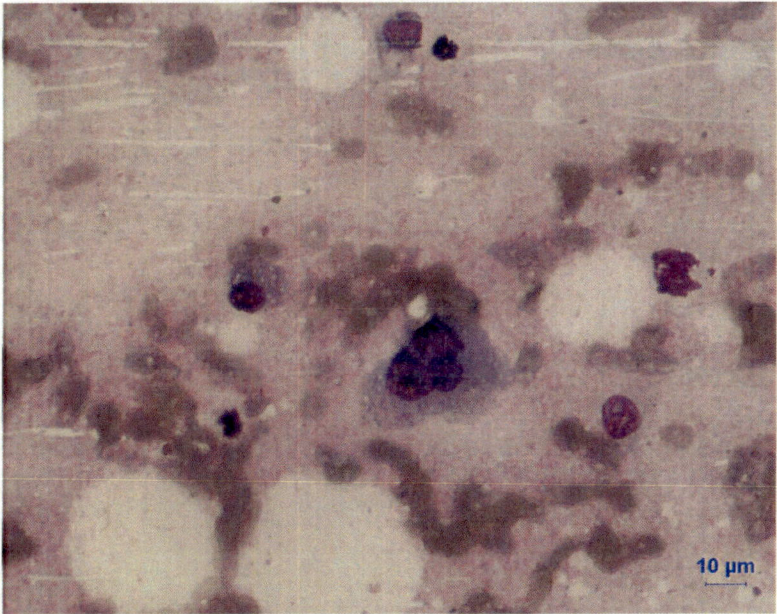

Osteosarcoma- Trinucleated cell Leishman & Giemsa Scale Bar 10 µm

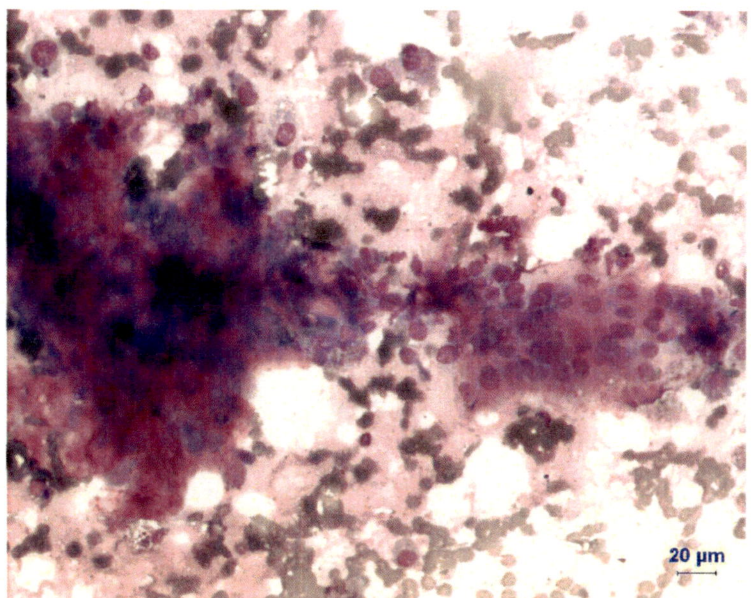

Osteosarcoma- Cluster of plasmacytoid cells with eosinophilic osteoid matrix
Leishman & Giemsa Scale Bar 20 µm

13

Cytological Diagnosis of Testicular Tumours

Testicular Tumours

Seminoma

- Cells were round exhibiting anisocytosis
- Round to oval nuclei
- Single distinct eosinophilic to basophilic nucleoli
- Cytoplasm scanty

Sertoli Cell Tumours

- Cellularity - High, anisocytosis
- Nucleus round to oval, megakaryosis, 1-3 basophilic nucleoli
- Chromatin coarse and granular
- Cytoplasm vacuolated

Leydig Cell Tumour/Interstitial Cell Tumour

- Cell clusters seen around capillary
- Anisocytosis
- Cytoplasm-Abundant, basophilic containing small vacuoles
- Occasional small black granules in a few cells seen
- N:Cy ratio low
- Nuclei-Small to medium sized; many contain nucleoli
- Chromatin-Fine reticular or homogeneous

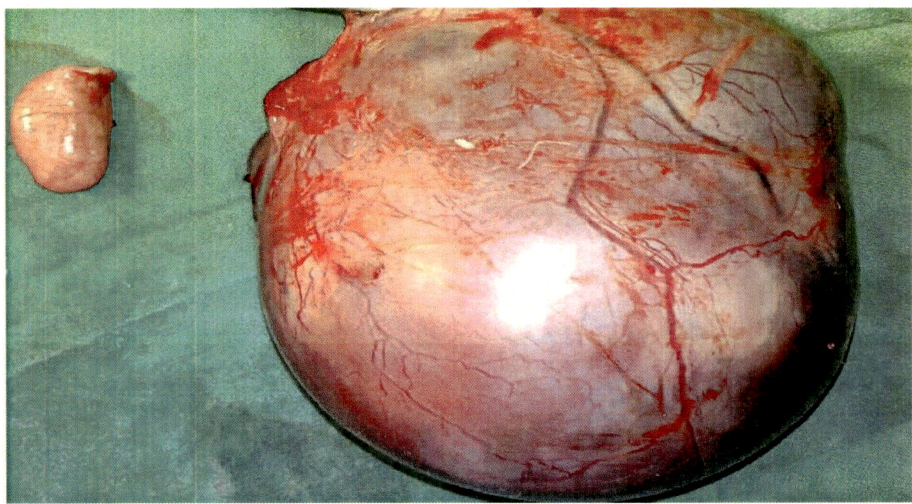

Dog-Seminoma-Enlarged cryptorchid testicle showed 5 cm diameter mass (Right) and normal testis (Left)

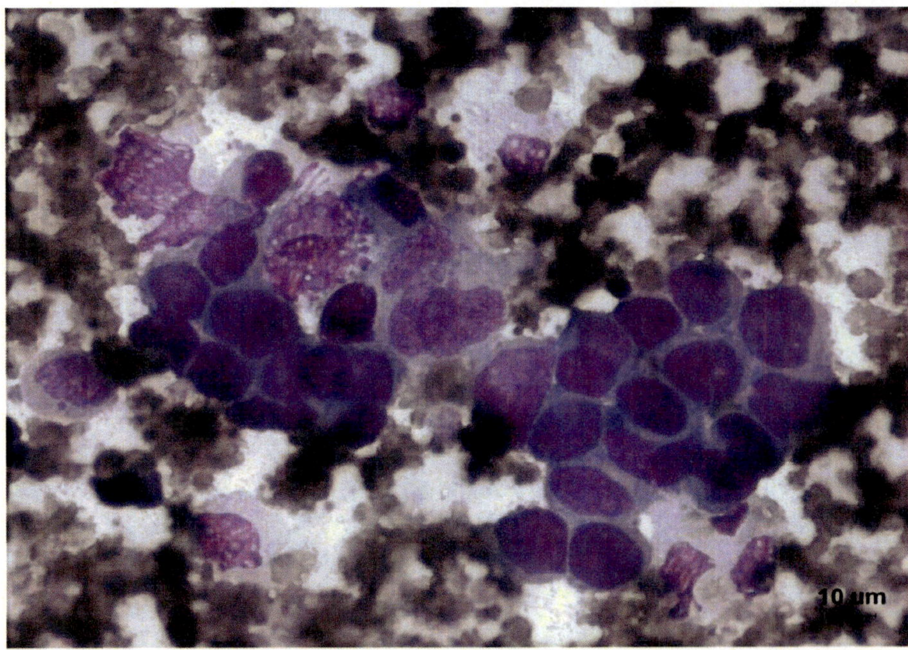

Dog-Seminoma- Large polyhedral shaped cells with vesicular nuclei and light basophilic cytoplasm. Prominent nucleoli and fine to coarse chromatin Leishman & Giemsa Scale Bar 10 µm

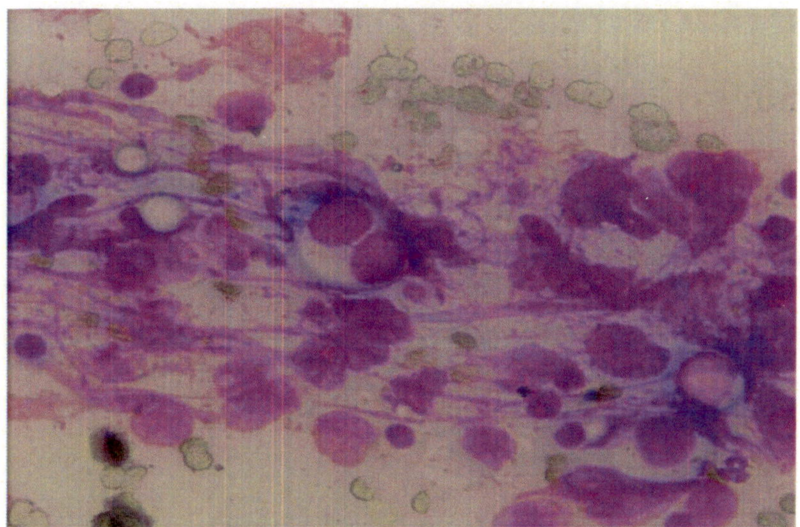

Dog-Seminoma-Large polygonal shaped cells with binucleated cell Leishman & Giemsa x 400

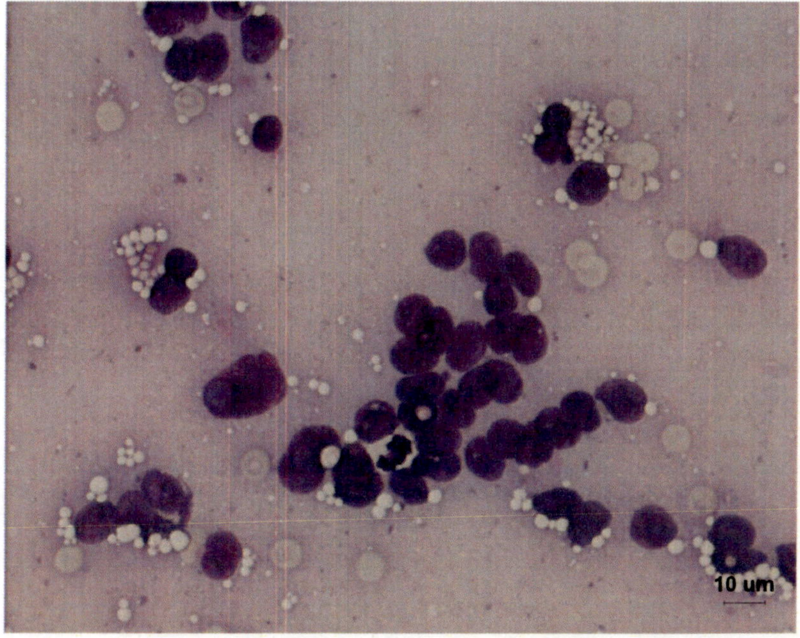

Dog- Sertoli cell tumour- Clusters of round cells with round to oval nuclei had coarse chromatin pattern and prominent nucleoli Cytoplasmic vacuolations are seen Leishman & Giemsa Scale Bar 10 µm

14

Cytological Diagnosis of Thyroid Tumours

Canine Thyroid Tumours

- Occurs as carcinoma
- Blood contamination also seen
- Cluster of cells
- Free nuclei seen in the pale-blue cytoplasm
- Dark blue to black pigment present in the cytoplasm
- Amorphous pink materials represents colloid
- Colloid and pigmented granuls indicates the thyroid origin
- Round to oval nuclei

Feline Thyroid Tumours

- It always occurs as adenoma
- Anisocytosis and anisokaryosis
- Binucleation
- Black granular cytoplasmic inclusions
- Cytopasm had poorly defined borders
- Pink colloid within the cluster
- Uniform nuclei

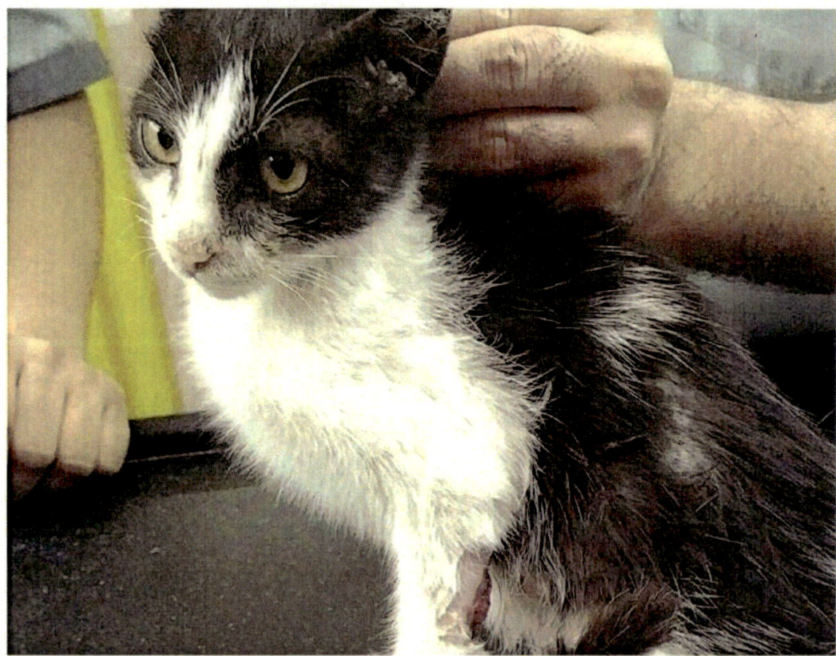

Thyroid adenoma -Cat- Swelling on the neck region (*Courtesy:*Dr. G. R. Baranidharan)

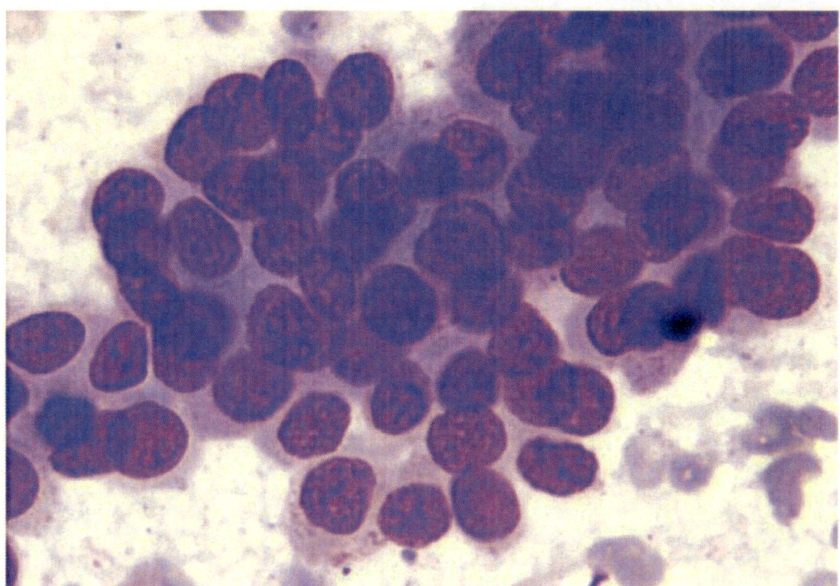

Thyroid adenoma -Clusters of thyroid follicular epithelial cells with round nuclei and moderate amount of basophilic cytoplasm Leishman & Giemsa x 400

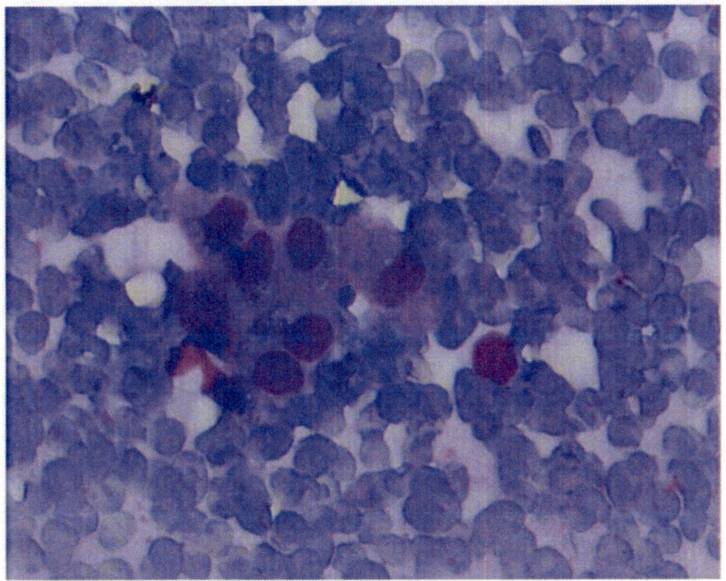

Thyroid adenoma- Presence of blue black granules in the cytoplasm
Leishman & Giemsa x 400

15

Cytological Diagnosis of Adrenal Tumours

Adrenal Tumours

- Adrenal cortical adenoma had normal secretory cells
- Most cells had naked nuclei in the abundant cytoplasm
- Basophilic cytoplasm with abundant clear lipid vacuoles
- Round uniform nuclei
- Prominent often single nucleolus

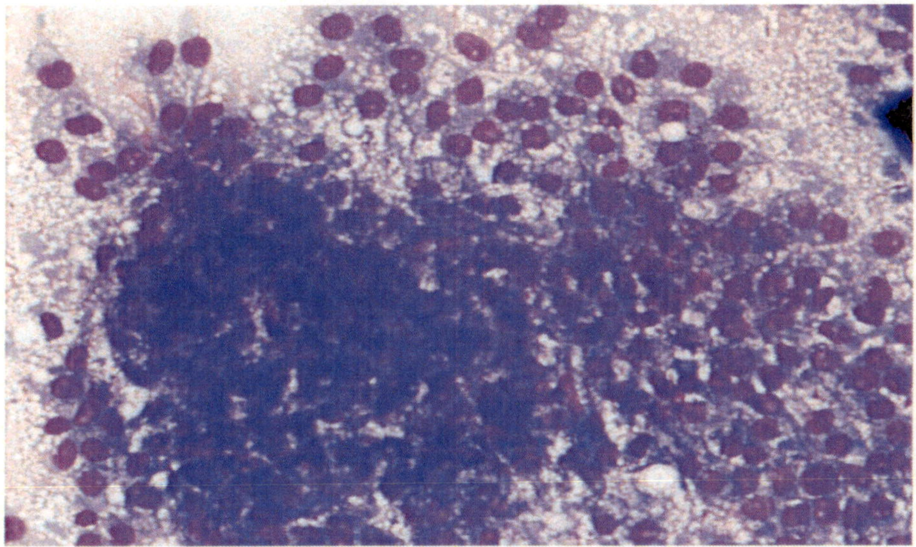

Dog-Adrenal cortical adenoma-Single to cluster of round to polygonal cells with round to oval nuclei and fine to coarse granular chromatin. Naked nuclei are also seen. Cytoplasm granular and vacuolated. Leishman & Giemsa x 200

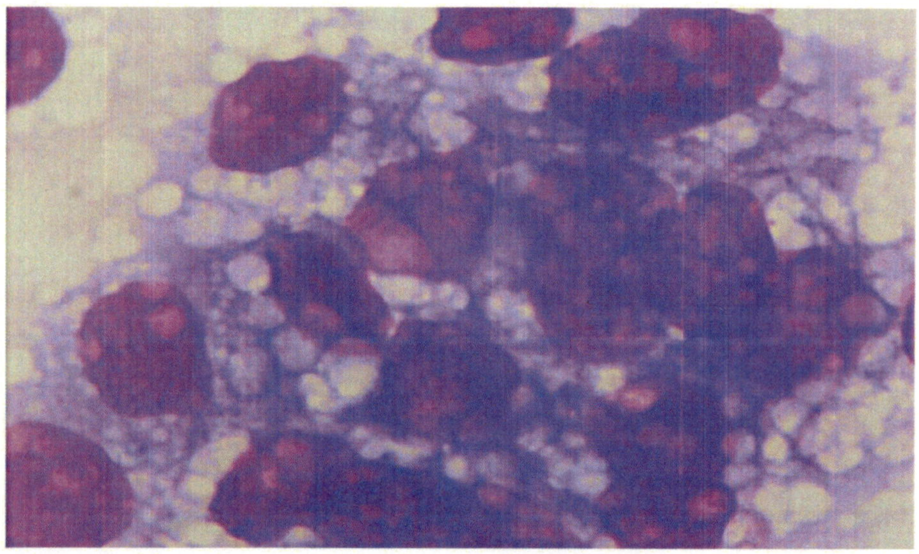

Dog-Adrenal cortical adenoma-Neoplastic round cells with round to oval nuclei and prominent cytoplasmic vacuolations. Leishman & Giemsa x 400

16

Cytological Diagnosis of Hepatic Tumours

Hepatocellular Carcinoma

- The presence of normal appearing hepatocytes of atypical hepatocytes with malignant characters
- Single to clusters of polyhedral or polygonal shaped or pleomorphic cells
- Large round, spherical or oval nuclei
- Higher nuclear cytoplasmic ratio
- Large nucleoli
- Dark blue cytoplasm

Cholangiocellular Carcinoma

- Sheets or clusters of cuboidal to low columnar cells
- Acinar or tubular formation may be seen
- Anisocytosis and anisockaryosis
- Round or oval nuclei
- Smooth finely reticulated chromatin
- Indistinct nucleoli
- Presence of black granules suggestive of bile accumulation

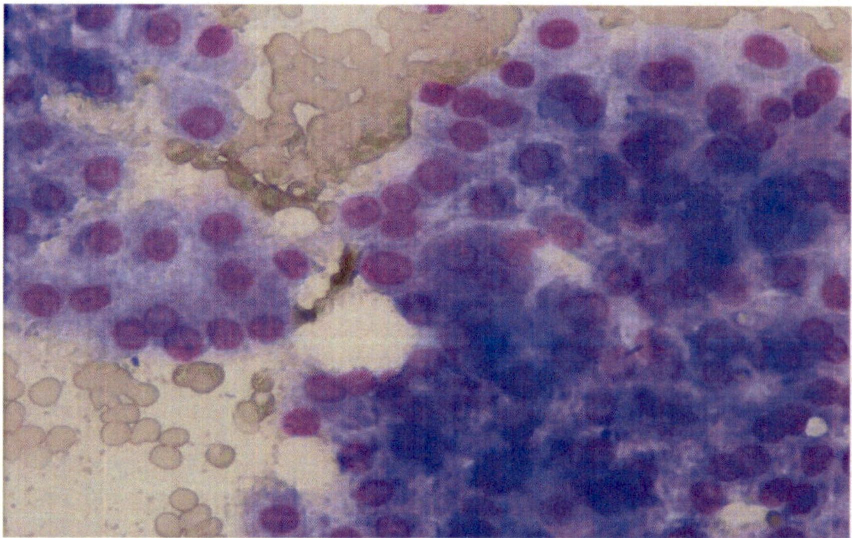

Hepatocellular carcinoma-Cluster of neoplastic plumpy polygonal cells with large round, spherical to oval nuclei and contained prominent nucleoli. Basophilic cytoplasm. Leishman & Giemsa x 400

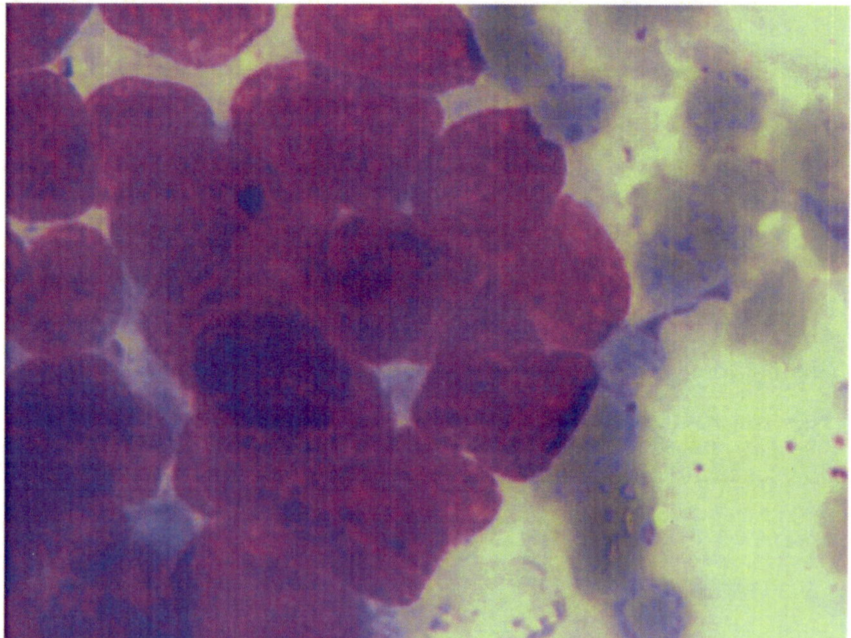

Hepatocellular carcinoma-Cluster of neoplastic plumpy polygonal cells with large round, spherical to oval nuclei. Leishman & Giemsa x 400

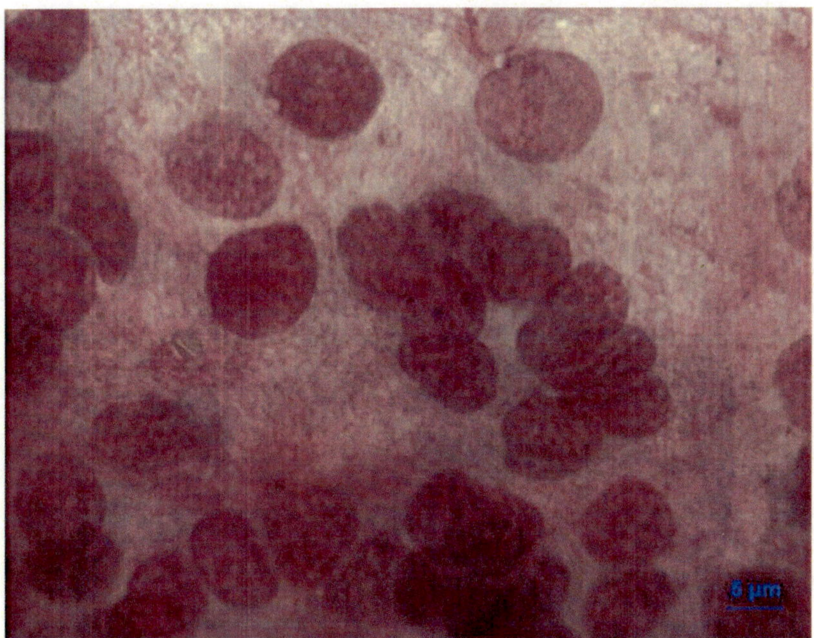

Cholangiocellular carcinoma- Cluster of cuboidal to low columnar cells arranged in acinar pattern. Nuclei are round to oval with fine to coarse chromatin and indistinct nucleoli. Leishman & Giemsa Scale Bar 5 µm

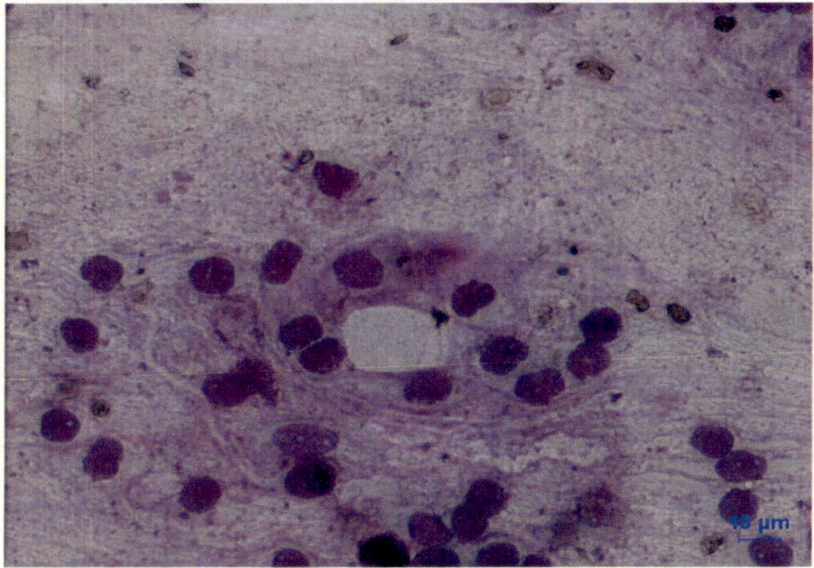

Cholangiocellular carcinoma- Neoplastic cells arranged in acinar pattern. Leishman & Giemsa Scale Bar 10 µm

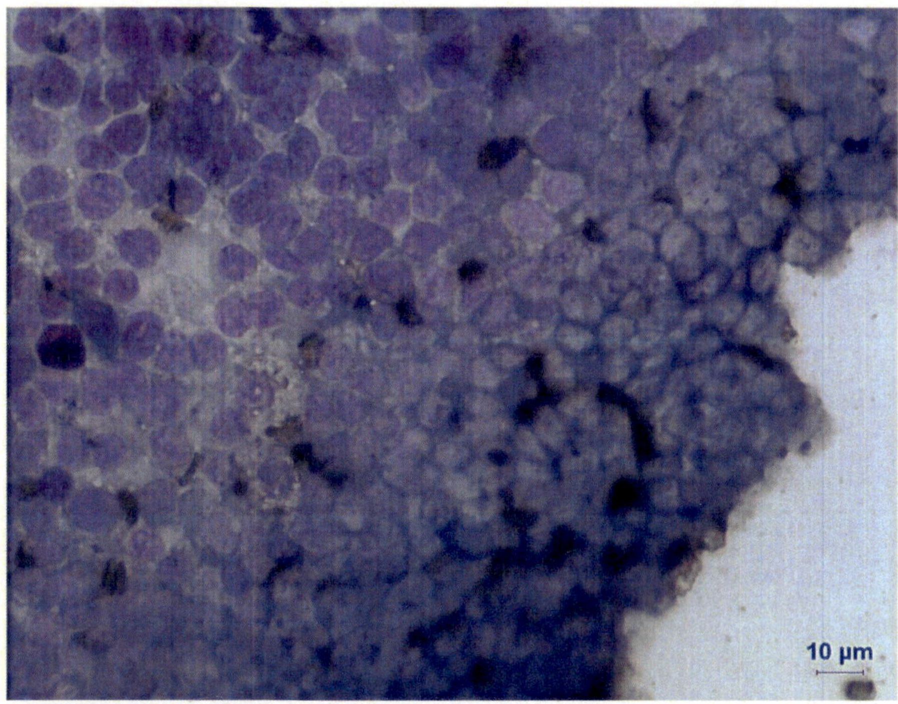

Cholangocellular carcinoma- Cholestasis- Presence of black granules suggestive of bile accumulation and bile casts filled the canaliculi between contiguous hepatocytes. Leishman & Giemsa Scale Bar 10 µm

17

Cytological Diagnosis of Mesothelioma

Mesothelioma

- Arising from the mesothelial lining of serous cavities
- It is very difficult to differentiate from carcinoma, reactive mesothelial hyperplasia
- Round to polygonal cells
- Arranged mostly in clusters
- Centrally located hyperchromatic nuclei
- Nuclear molding

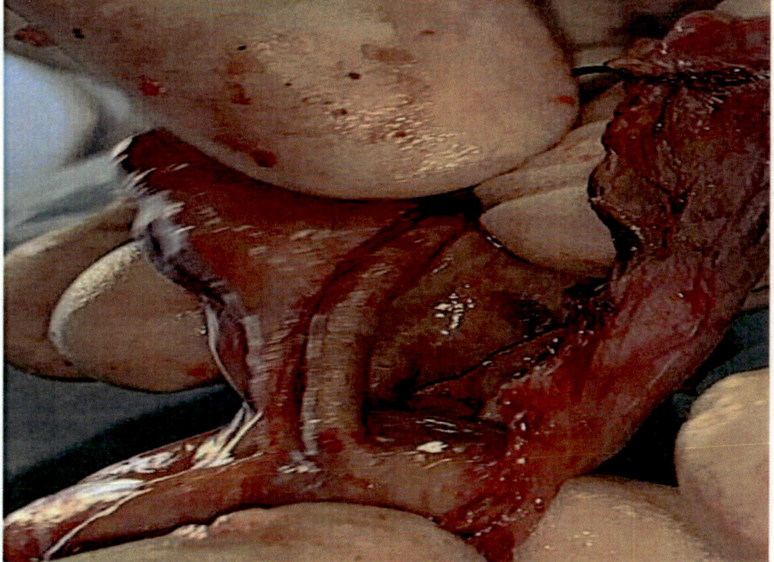

Mesothelioma-Pericardium- Thickened with minute gray white areas

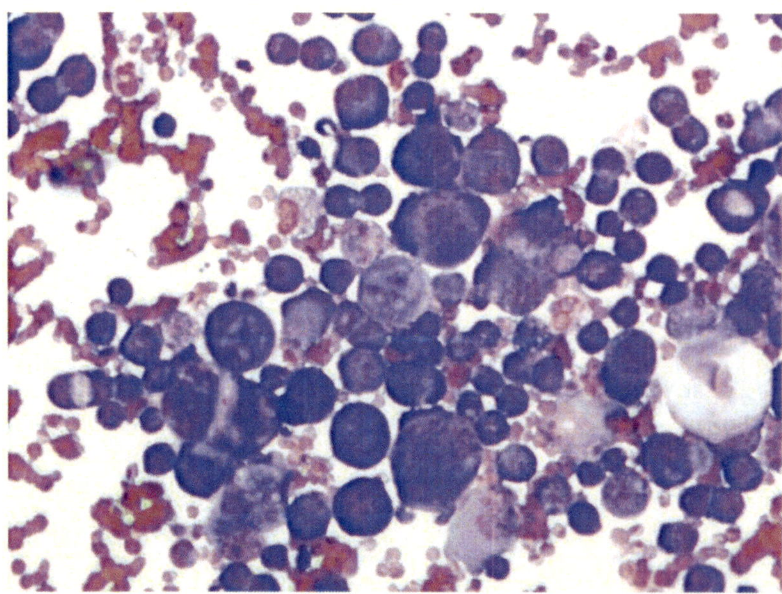

Dog-Mesothelioma- Single to small cluster of mesothelial cells with bi, tri and multinucleation. Leishman & Giemsa x 200

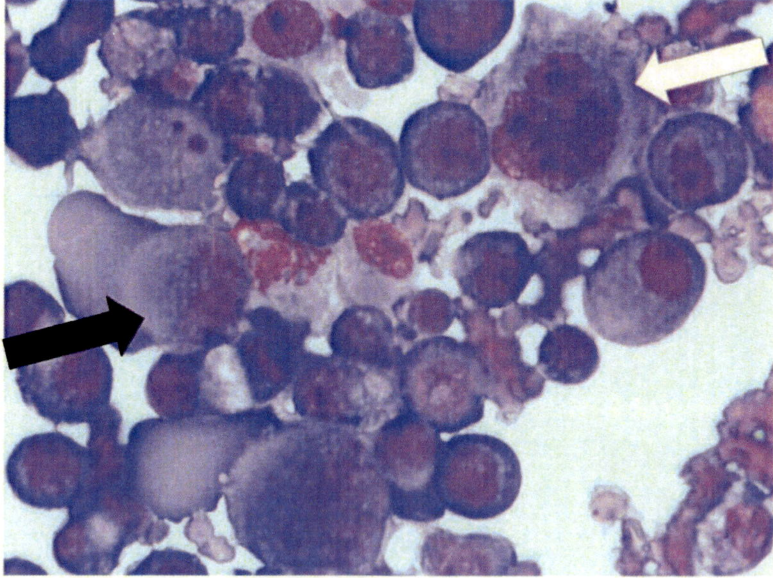

Dog-Mesothelioma- Cell in cell interaction (black arrow) and multinucleated cells (white arrow). Leishman & Giemsa x 200

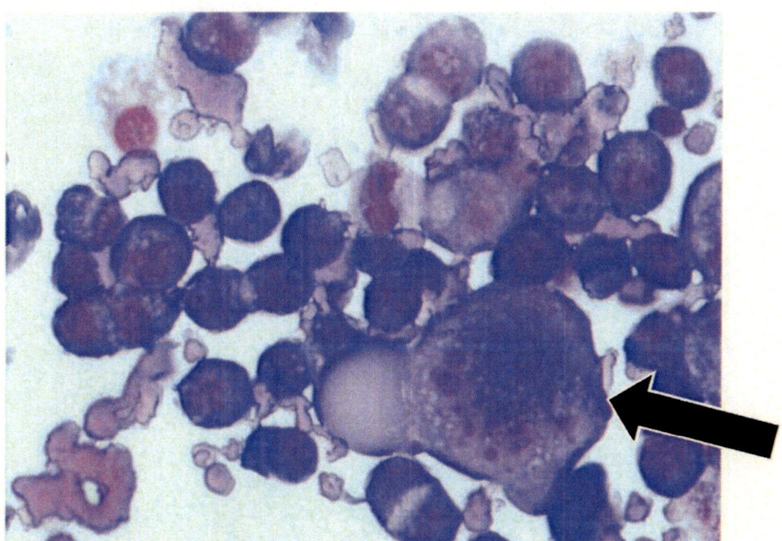

Dog-Mesothelioma- Giant cell (arrow) with pink endoplasm and blue ectoplasm. Leishman & Giemsa x 200

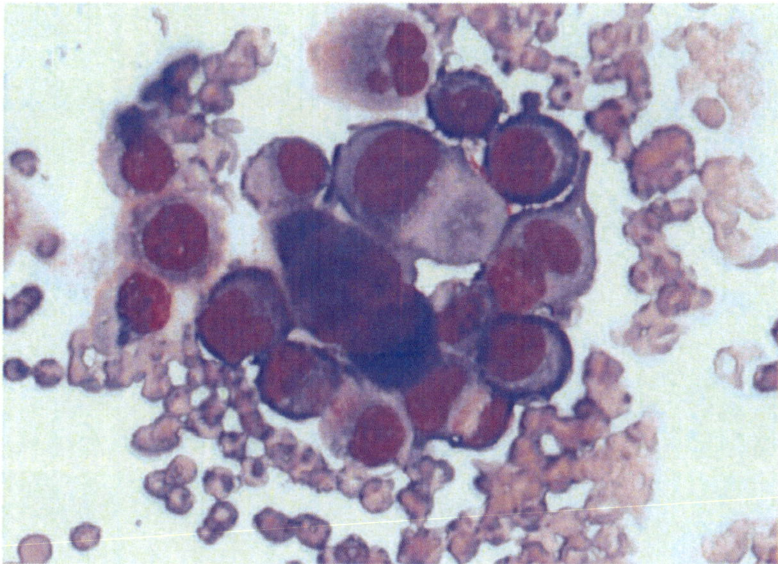

Dog-Mesothelioma- Small cluster of mesothelial cells. Binucleated cell is seen. . Leishman & Giemsa x 200

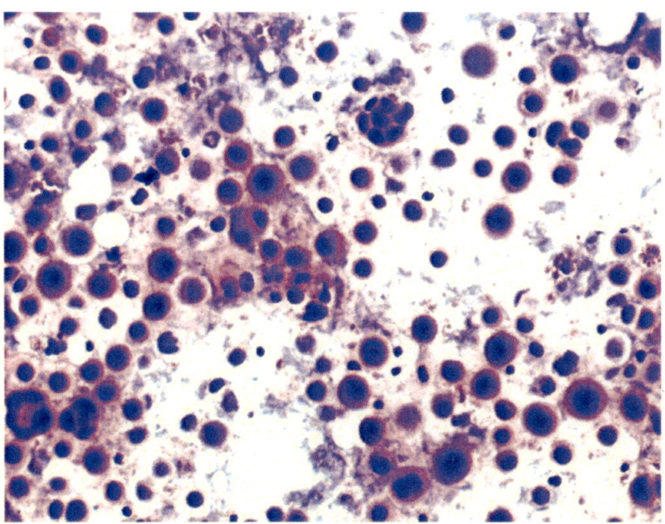

Pericardial fluid- Mesothelioma - Mesothelioma - Single to cluster of epithelioid neoplastic mesothelial cells and knobby appearance with eosinophilic cytoplasm. Binucleate cell. Leishman & Giemsa x 200

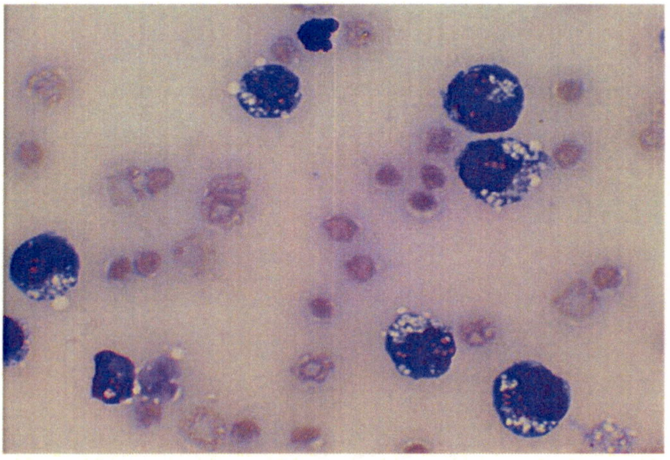

Pericardial fluid-Macrophages. Vacuolated cytoplasm. Leishman & Giemsa x 200

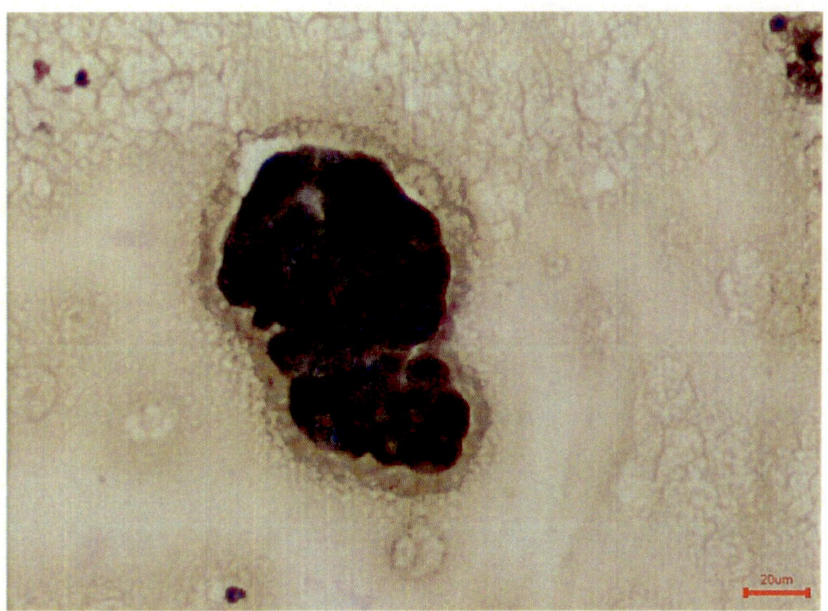

Dog-Mesothelioma-Immunohistochemistry-Vimentin-Cytoplasmic expression of mesothelial cells. Scale Bar 20 μm

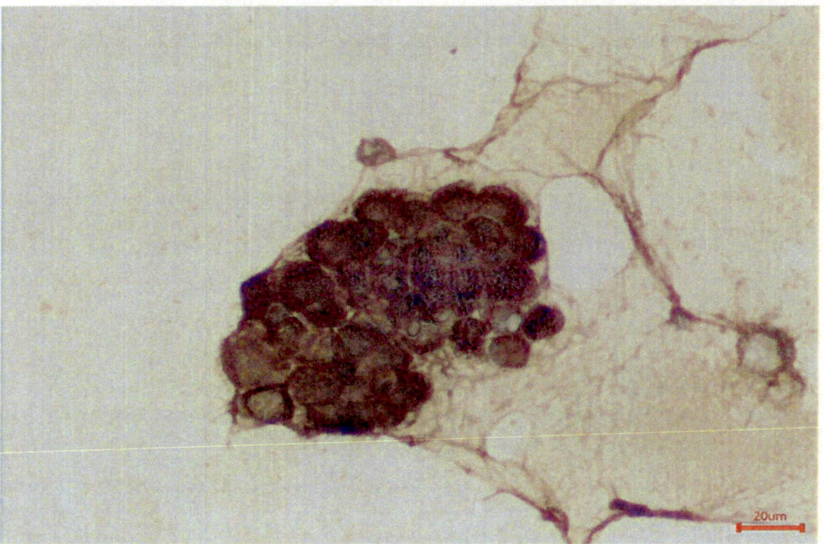

Dog-Mesothelioma- Cytokeratin-Cytoplasmic expression of mesothelial cells. Scale Bar 20 μm

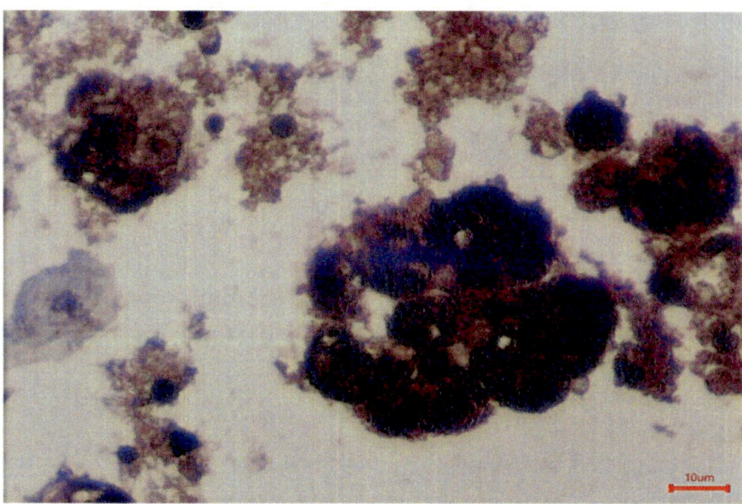

Dog-Mesothelioma- Mesothelin-Membrane expression of mesothelial cells.Scale Bar 10 μm

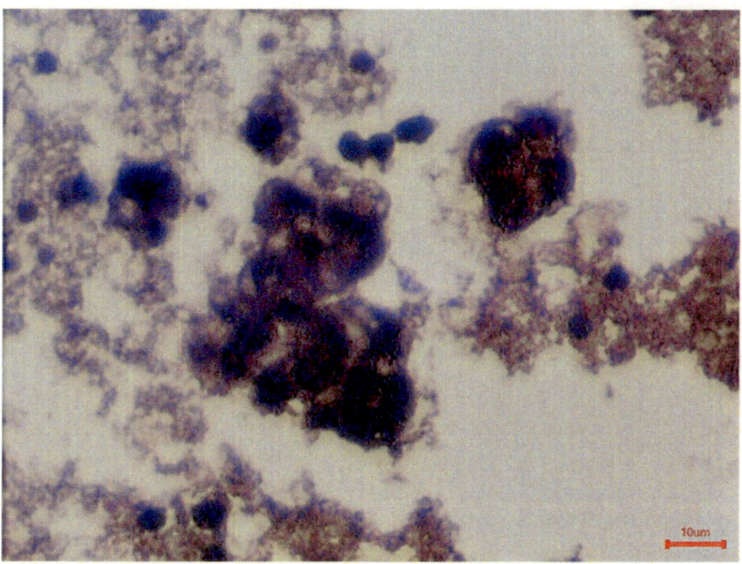

Dog-Mesothelioma- Mesothelin-Membrane expression of mesothelial cells Scale Bar 10 μm

References

Burton, A.G.2018. Clinical Atlas of Small Animal Cytology. John Wiley and Sons Inc, New Jersey.

Cowell,R.L., Tyler,R.D.,Meinkoth,J.H. and DeNicola,D.B. (2008).Diagnostic

Cytology and Hematology of the Dog and Cat.Third Edition. Mosby Elsevier.

Dunn,J.(2014). Manual of Diagnostic Cytology of the Dog and Cat . I[st] Edn. Wiley-Blackwell

Lumsden, J.H. and R. Baker. (2000). Principle of Cytological Evaluation and Cytological Technique and Interpretation. In: Colour Atlas of Cytology of The Dog and Cat. I[st]Edn. Mobsy, pp. 3-20

Krithiga, K., Murali Manohar, B., & Balachandran, C. (2005). Cytological and histopathological diagnosis of canine skin and adnexae cell tumours.Indian J. Vet. Pathol., 29(2), 112-117.

Krithiga, K., Murali Manohar, B., & Balachandran, C. (2005). Cytological and histopathollogical diagnosis. I. Canine mammary tumours. Indian J. Vet. Pathol., 29(2), 118- 120.

Krithiga, K., Murali Manohar, B., & Balachandran, C. (2005). Cytological and histopathological diagnosis. II. Canine mesenchymal tumours. Indian J. Vet. Pathol., 29(2), 121-124.

Meinkoth, J.H. and R.L. Cowell. (2002). Sample Collection and Preparation in Cytology: Increasing Diagnostic Yield. Vet. Clin. Small Anim., 32: 1187-1207

Moulton, J.E. 1990. Tumours in Domestic Animals. 3rd edn., Ed: J.E.Moulton, University of California Press, London. pp: 231-307.

Roszel, J.F. (1981). Cytological procedure. J. Am. Anim. Hosp. Assoc., 17: 903-910

Sangha, S., Amarjit Singh, Sood, N.K. and Kuldip Gupta. (2011). Specificity and sensitivity of cytological techniques for rapid diagnosis of neoplastic and non-neoplastic lesions of canine mammary gland. Braz. J. Vet. Pathol., 4(1), 13-22

Stirtzinger, T. (1988). The Cytologic diagnosis of mesenchymal tumors. Semin. Vet. Med. Surg. (Small Anim)., 3: 151

Priya, S. 2005. Comparison of utility of nipple aspirates fluid cytology with other routine diagnostic tests. M.V.Sc., thesis approved by to TANUVAS, Chennai - 600 051

Raskin, R.E. and Meyer, D.J. (2010). Canine and feline cytology: A Colour Atlas and Interpretation Guide. Saunders Elsevier. Second Edn. Missouri, USA.

Thangapandiyan, M. and Balachnadran, C. (2010). Cytological evaluation of canine Lymphadenopathies -a review of 109 cases. 80 (4):499-508

Thangapandiyan, M. (2015). Cytohistological and immunocytochemical studies on lymphadenopathies with special reference to immunophenotyping of lymphoma in canines. Ph.D., thesis approved by TANUVAS, Chennai-600051. India

Index